storytelling
with data

a data visualization guide
for business professionals

cole nussbaumer knaflic

WILEY

Cover image: Cole Nussbaumer Knaflic
Cover design: Wiley

Published by John Wiley & Sons, Inc., Hoboken, New Jersey.

Published simultaneously in Canada.

For general information on our other products and services or for technical support, please contact our Customer Care Department within the United States at (800) 762-2974, outside the United States at (317) 572-3993 or fax (317) 572-4002.

Wiley publishes in a variety of print and electronic formats and by print-on-demand. Some material included with standard print versions of this book may not be included in e-books or in print-on-demand. If this book refers to media such as a CD or DVD that is not included in the version you purchased, you may download this material at http://booksupport.wiley.com. For more information about Wiley products, visit www.wiley.com.

Library of Congress Cataloging-in-Publication Data:

ISBN 9781119002253 (Paperback)
ISBN 9781119002260 (ePDF)
ISBN 9781119002062 (ePub)

Printed in the United States of America

SKY10022002_102920

To Randolph

contents

foreword   ix

acknowledgments   xi

about the author   xiii

introduction   1

chapter 1   the importance of context   19

chapter 2   choosing an effective visual   35

chapter 3   clutter is your enemy!   71

chapter 4   focus your audience's attention   99

chapter 5   think like a designer   127

chapter 6   dissecting model visuals   151

chapter 7   lessons in storytelling   165

chapter 8   pulling it all together   187

chapter 9   case studies   207

chapter 10   final thoughts   241

bibliography   257

index   261

foreword

"Power Corrupts. PowerPoint Corrupts Absolutely."

—*Edward Tufte, Yale Professor Emeritus*[1]

We've all been victims of bad slideware. Hit-and-run presentations that leave us staggering from a maelstrom of fonts, colors, bullets, and highlights. Infographics that fail to be informative and are only graphic in the same sense that violence can be graphic. Charts and tables in the press that mislead and confuse.

It's too easy today to generate tables, charts, graphs. I can imagine some old-timer (maybe it's me?) harrumphing over my shoulder that in *his* day they'd do illustrations by hand, which meant you had to *think* before committing pen to paper.

Having all the information in the world at our fingertips doesn't make it easier to communicate: it makes it harder. The more information you're dealing with, the more difficult it is to filter down to the most important bits.

Enter Cole Nussbaumer Knaflic.

I met Cole in late 2007. I'd been recruited by Google the year before to create the "People Operations" team, responsible for finding, keeping, and delighting the folks at Google. Shortly after joining I decided

---

[1] Tufte, Edward R. 'PowerPoint Is Evil.' Wired Magazine, www.wired.com/wired/archive/11.09/ppt2.html, September 2003.

we needed a People Analytics team, with a mandate to make sure we innovated as much on the people side as we did on the product side. Cole became an early and critical member of that team, acting as a conduit between the Analytics team and other parts of Google.

Cole always had a knack for clarity.

She was given some of our messiest messages—such as what exactly makes one manager great and another crummy—and distilled them into crisp, pleasing imagery that told an irrefutable story. Her messages of "don't be a data fashion victim" (i.e., lose the fancy clipart, graphics and fonts—focus on the message) and "simple beats sexy" (i.e., the point is to clearly tell a story, not to make a pretty chart) were powerful guides.

We put Cole on the road, teaching her own data visualization course over 50 times in the ensuing six years, before she decided to strike out on her own on a self-proclaimed mission to "rid the world of bad PowerPoint slides." And if you think that's not a big issue, a Google search of "powerpoint kills" returns almost half a million hits!

In *Storytelling with Data*, Cole has created an of-the-moment complement to the work of data visualization pioneers like Edward Tufte. She's worked at and with some of the most data-driven organizations on the planet as well as some of the most mission-driven, data-free institutions. In both cases, she's helped sharpen their messages, and their thinking.

She's written a fun, accessible, and eminently practical guide to extracting the signal from the noise, and for making all of us better at getting our voices heard.

And that's kind of the whole point, isn't it?

Laszlo Bock

SVP of People Operations, Google, Inc.
and author of *Work Rules!*

May 2015

acknowledgments

My timeline of thanks

Thank you to...

2015

1980

2010–CURRENT **My family**, for your love and support. To my love, my husband, Randy, for being my #1 cheerleader through it all; I love you, darling. To my beautiful sons, Avery and Dorian, for reprioritizing my life and bringing much joy to my world.

2010–CURRENT **My clients**, for taking part in my effort to rid the world of ineffective graphs and inviting me to share my work with their teams and organizations through workshops and other projects.

2007–2012 **The Google Years**. Laszlo Bock, Prasad Setty, Brian Ong, Neal Patel, Tina Malm, Jennifer Kurkoski, David Hoffman, Danny Cohen, and Natalie Johnson, for giving me the opportunity and autonomy to research, build, and teach content on effective data visualization, for subjecting your work to my often critical eye, and for general support and inspiration.

2002–2007 **The Banking Years**. Mark Hillis and Alan Newstead, for recognizing and encouraging excellence in visual design as I first started to discover and hone my data viz skills (in sometimes painful ways, like the fraud management spider graph!).

1987–CURRENT **My brother**, for reminding me of the importance of balance in life.

1980–CURRENT **My dad**, for your design eye and attention to detail.
1980–2011 **My mother**, the single biggest influence on my life; I miss you, Mom.

Thank you also to everyone who helped make this book possible. I value every bit of input and help along the way. In addition to the people listed above, thanks to Bill Falloon, Meg Freeborn, Vincent Nordhaus, Robin Factor, Mark Bergeron, Mike Henton, Chris Wallace, Nick Wehrkamp, Mike Freeland, Melissa Connors, Heather Dunphy, Sharon Polese, Andrea Price, Laura Gachko, David Pugh, Marika Rohn, Robert Kosara, Andy Kriebel, John Kania, Eleanor Bell, Alberto Cairo, Nancy Duarte, Michael Eskin, Kathrin Stengel, and Zaira Basanez.

about the author

Cole Nussbaumer Knaflic tells stories with data. She specializes in the effective display of quantitative information and writes the popular blog storytellingwithdata.com. Her well-regarded workshops and presentations are highly sought after by data-minded individuals, companies, and philanthropic organizations all over the world.

Her unique talent was honed over the past decade through analytical roles in banking, private equity, and most recently as a manager on the Google People Analytics team. At Google, she used a data-driven approach to inform innovative people programs and management practices, ensuring that Google attracted, developed, and retained great talent and that the organization was best aligned to meet business needs. Cole traveled to Google offices throughout the United States and Europe to teach the course she developed on data visualization. She has also acted as an adjunct faculty member at the Maryland Institute College of Art (MICA), where she taught Introduction to Information Visualization.

Cole has a BS in Applied Math and an MBA, both from the University of Washington. When she isn't ridding the world of ineffective graphs one pie at a time, she is baking them, traveling, and embarking on adventures with her husband and two young sons in San Francisco.

introduction

Bad graphs are everywhere

I encounter a lot of less-than-stellar visuals in my work (and in my life—once you get a discerning eye for this stuff, it's hard to turn it off). Nobody sets out to make a bad graph. But it happens. Again and again. At every company throughout all industries and by all types of people. It happens in the media. It happens in places where you would expect people to know better. Why is that?

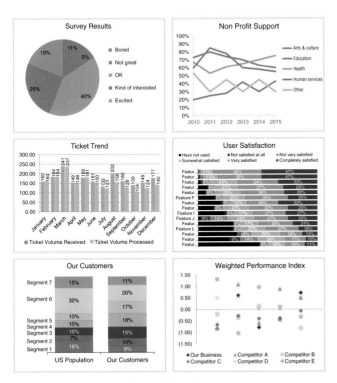

FIGURE 0.1   A sampling of ineffective graphs

We aren't naturally good at storytelling with data

In school, we learn a lot about language and math. On the language side, we learn how to put words together into sentences and into stories. With math, we learn to make sense of numbers. But it's rare that these two sides are paired: no one teaches us how to tell stories with numbers. Adding to the challenge, very few people feel naturally adept in this space.

This leaves us poorly prepared for an important task that is increasingly in demand. Technology has enabled us to amass greater and greater amounts of data and there is an accompanying growing desire to make sense out of all of this data. Being able to visualize data and tell stories with it is key to turning it into *information* that can be used to drive better decision making.

In the absence of natural skills or training in this space, we often end up relying on our tools to understand best practices. Advances in technology, in addition to increasing the amount of and access to data, have also made tools to work with data pervasive. Pretty much anyone can put some data into a graphing application (for example, Excel) and create a graph. This is important to consider, so I will repeat myself: *anyone* can put some data into a graphing application and create a graph. This is remarkable, considering that the process of creating a graph was historically reserved for scientists or those in other highly technical roles. And scary, because without a clear path to follow, our best intentions and efforts (combined with oft-questionable tool defaults) can lead us in some really bad directions: 3D, meaningless color, pie charts.

Skilled in Microsoft Office? So is everyone else!

Being adept with word processing applications, spread-sheets, and presentation software—things that used to set one apart on a resume and in the workplace—has become a minimum expectation for most employers. A recruiter told me that, today, having "proficiency in Microsoft Office" on a resume isn't enough: a basic level of knowledge here is assumed and it's what you can do above and beyond that will set you apart from others. Being able to effectively tell stories with data is one area that will give you that edge and position you for success in nearly any role.

While technology has increased access to and proficiency in tools to work with data, there remain gaps in capabilities. You can put some data in Excel and create a graph. For many, the process of data visualization ends there. This can render the most interesting story completely underwhelming, or worse—difficult or impossible to understand. Tool defaults and general practices tend to leave our data and the stories we want to tell with that data sorely lacking.

There is a story in your data. But your tools don't know what that story is. That's where it takes you—the analyst or communicator of the information—to bring that story visually and contextually to life. That process is the focus of this book. The following are a few example before-and-afters to give you a visual sense of what you'll learn; we'll cover each of these in detail at various points in the book.

The lessons we will cover will enable you to shift from simply showing data to **storytelling with data.**

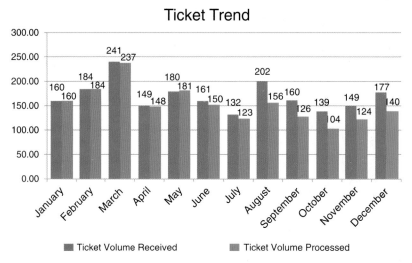

FIGURE 0.2   Example 1 (before): showing data

Please approve the hire of 2 FTEs
to backfill those who quit in the past year

Ticket volume over time

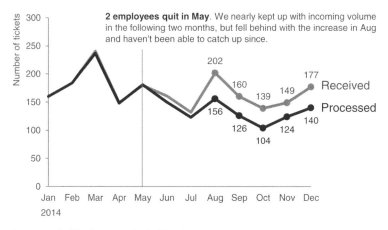

Data source: XYZ Dashboard, as of 12/31/2014 | A detailed analysis on tickets processed per person and time to resolve issues was undertaken to inform this request and can be provided if needed.

FIGURE 0.3   Example 1 (after): storytelling with data

Survey Results

PRE: How do you feel about doing science?

■ Bored ■ Not great ■ OK ■ Kind of interested ■ Excited

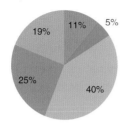

POST: How do you feel about doing science?

■ Bored ■ Not great ■ OK ■ Kind of interested ■ Excited

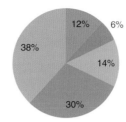

FIGURE 0.4   Example 2 (before): showing data

Pilot program was a success

How do you feel about science?

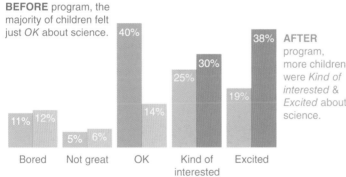

BEFORE program, the majority of children felt just *OK* about science.

AFTER program, more children were *Kind of interested* & *Excited* about science.

Based on survey of 100 students conducted before and after pilot program (100% response rate on both surveys).

FIGURE 0.5   Example 2 (after): storytelling with data

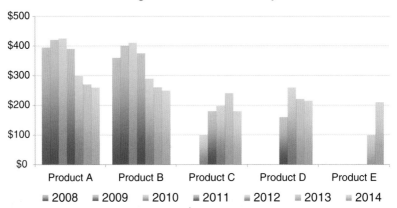

FIGURE 0.6   Example 3 (before): showing data

To be competitive, we recommend introducing our product *below the $223 average* price point in the **$150–$200 range**

Retail price over time by product

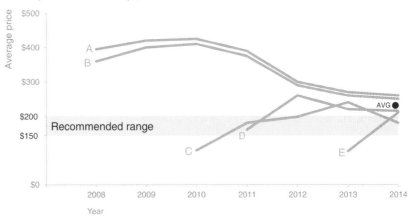

FIGURE 0.7   Example 3 (after): storytelling with data

Who this book is written for

This book is written for anyone who needs to communicate some-*thing* to some*one* using data. This includes (but is certainly not limited to): analysts sharing the results of their work, students visualizing thesis data, managers needing to communicate in a data-driven way, philanthropists proving their impact, and leaders informing their board. I believe that anyone can improve their ability to communicate effectively with data. This is an intimidating space for many, but it does not need to be.

When you are asked to "show data," what sort of feelings does that evoke?

Perhaps you feel uncomfortable because you are unsure where to start. Or maybe it feels like an overwhelming task because you assume that what you are creating needs to be complicated and show enough detail to answer every possible question. Or perhaps you already have a solid foundation here, but are looking for that something that will help take your graphs and the stories you want to tell with them to the next level. In all of these cases, this book is written with you in mind.

"When I'm asked to *show the data*, I feel..."

An informal Twitter poll I conducted revealed the following mix of emotions when people are asked to "show the data."

Frustrated because I don't think I'll be able to tell the whole story.

Pressure to make it clear to whomever needs the data.

Inadequate. Boss: Can you drill down into that? Give me the split by x, y, and z.

Being able to tell stories with data is a skill that's becoming ever more important in our world of increasing data and desire for data-driven decision making. An effective data visualization can mean the difference between success and failure when it comes to communicating the findings of your study, raising money for your non-profit, presenting to your board, or simply getting your point across to your audience.

My experience has taught me that most people face a similar challenge: they may recognize the need to be able to communicate effectively with data but feel like they lack expertise in this space. People skilled in data visualization are hard to come by. Part of the challenge is that data visualization is a single step in the analytical process. Those hired into analytical roles typically have quantitative backgrounds that suit them well for the other steps (finding the data, pulling it together, analyzing it, building models), but not necessarily any formal training in design to help them when it comes to the communication of the analysis—which, by the way, is typically the only part of the analytical process that your audience ever sees. And increasingly, in our ever more data-driven world, those without technical backgrounds are being asked to put on analytical hats and communicate using data.

The feelings of discomfort you may experience in this space aren't surprising, given that being able to communicate effectively with data isn't something that has been traditionally taught. Those who excel have typically learned what works and what doesn't through trial and error. This can be a long and tedious process. Through this book, I hope to help expedite it for you.

How I learned to tell stories with data

I have always been drawn to the space where mathematics and business intersect. My educational background is mathematics and business, which enables me to communicate effectively with both sides—given that they don't always speak the same language—and help them better understand one another. I love being able to take the science of data and use it to inform better business decisions. Over time, I've found that one key to success is being able to communicate effectively visually with data.

I initially recognized the importance of being skilled in this area during my first job out of college. I was working as an analyst in credit risk management (before the subprime crisis and hence before anyone really knew what credit risk management was). My job was to build and assess statistical models to forecast delinquency and loss. This meant taking complicated stuff and ultimately turning it into a simple communication of whether we had adequate money in the reserves for expected losses, in what scenarios we'd be at risk, and so forth. I quickly learned that spending time on the aesthetic piece—something my colleagues didn't typically do—meant my work garnered more attention from my boss and my boss's boss. For me, that was the beginning of seeing value in spending time on the visual communication of data.

After progressing through various roles in credit risk, fraud, and operations management, followed by some time in the private equity world, I decided I wanted to continue my career outside of banking and finance. I paused to reflect on the skills I possessed that I wanted to be utilizing on a daily basis: at the core, it was using data to influence business decisions.

I landed at Google, on the People Analytics team. Google is a data-driven company—so much so that they even use data and analytics in a space not frequently seen: human resources. People Analytics is an analytics team embedded in Google's HR organization (referred to at Google as "People Operations"). The mantra of this team is to help ensure that people decisions at Google—decisions about employees or future employees—are data driven. This was an amazing place to continue to hone my storytelling with data skills, using data and analytics to better understand and inform decision making in spaces like targeted hiring, engaging and motivating employees, building effective teams, and retaining talent. Google People Analytics is cutting edge, helping to forge a path that many other companies have started to follow. Being involved in building and growing this team was an incredible experience.

Storytelling with data on what makes a great manager via Project Oxygen

One particular project that has been highlighted in the public sphere is the Project Oxygen research at Google on what makes a great manager. This work has been described in the *New York Times* and is the basis of a popular *Harvard Business Review* case study. One challenge faced was communicating the findings to various audiences, from engineers who were sometimes skeptical on methodology and wanted to dig into the details, to managers wanting to understand the big-picture findings and how to put them to use. My involvement in the project was on the communication piece, helping to determine how to best show sometimes very complicated stuff in a way that would appease the engineers and their desire for detail while still being understandable and straightforward for managers and various levels of leadership. To do this, I leveraged many of the concepts we will discuss in this book.

The big turning point for me happened when we were building an internal training program within People Operations at Google and I was asked to develop content on data visualization. This gave me the opportunity to research and start to learn the principles behind effective data visualization, helping me understand why some of the things I'd arrived at through trial and error over the years had been effective. With this research, I developed a course on data visualization that was eventually rolled out to all of Google.

The course created some buzz, both inside and outside of Google. Through a series of fortuitous events, I received invitations to speak at a couple of philanthropic organizations and events on the topic of data visualization. Word spread. More and more people were reaching out to me—initially in the philanthropic world, but increasingly in the corporate sector as well—looking for guidance on how to communicate effectively with data. It was becoming increasingly clear that the need in this space was not unique to Google. Rather, pretty much anyone in an organization or business setting could increase their impact by being able to communicate effectively with data. After acting as a speaker at conferences and organizations in my spare time, eventually I left Google to pursue my emerging goal of teaching the world how to tell stories with data.

Over the past few years, I've taught workshops for more than a hundred organizations in the United States and Europe. It's been interesting to see that the need for skills in this space spans many industries and roles. I've had audiences in consulting, consumer products, education, financial services, government, health care, nonprofit, retail, startups, and technology. My audiences have been a mix of roles and levels: from analysts who work with data on a daily basis to those in non-analytical roles who occasionally have to incorporate data into their work, to managers needing to provide guidance and feedback, to the executive team delivering quarterly results to the board.

Through this work, I've been exposed to many diverse data visualization challenges. I have come to realize that the skills that are needed in this area are fundamental. They are not specific to any industry or role, and they can be effectively taught and learned—as demonstrated by the consistent positive feedback and follow-ups I receive from workshop attendees. Over time, I've codified the lessons that I teach in my workshops. These are the lessons I will share with you.

How you'll learn to tell stories with data: 6 lessons

In my workshops, I typically focus on five key lessons. The big opportunity with this book is that there isn't a time limit (in the way there is in a workshop setting). I've included a sixth bonus lesson that I've always wanted to share ("think like a designer") and also a lot more by way of before-and-after examples, step-by-step instruction, and insight into my thought process when it comes to the visual design of information.

I will give you practical guidance that you can begin using immediately to better communicate visually with data. We'll cover content to help you learn and be comfortable employing six key lessons:

1. Understand the context

2. Choose an appropriate visual display

3. Eliminate clutter

4. Focus attention where you want it

5. Think like a designer

6. Tell a story

Illustrative examples span many industries

Throughout the book, I use a number of case studies to illustrate the concepts discussed. The lessons we cover will not be industry—or role—specific, but rather will focus on fundamental concepts and best practices for effective communication with data. Because my work spans many industries, so do the examples upon which I draw. You will see case studies from technology, education, consumer products, the nonprofit sector, and more.

Each example used is based on a lesson I have taught in my workshops, but in many cases I've slightly changed the data or generalized the situation to protect confidential information.

For any example that doesn't initially seem relevant to you, I encourage you to pause and think about what data visualization or communication challenges you encounter where a similar approach could be effective. There is something to be learned from every example, even if the example itself isn't obviously related to the world in which you work.

Lessons are not tool specific

The lessons we will cover in this book focus on best practices that can be applied in any graphing application or presentation software. There are a vast number of tools that can be leveraged to tell effective stories with data. No matter how great the tool, however, it will never know your data and your story like you do. Take the time to learn your tool well so that it does not become a limiting factor when it comes to applying the lessons we'll cover throughout this book.

How do you do that in Excel?

While I will not focus the discussion on specific tools, the examples in this book were created using Microsoft Excel. For those interested in a closer look at how similar visuals can be built in Excel, please visit my blog at storytellingwithdata.com, where you can download the Excel files that accompany my posts.

How this book is organized

This book is organized into a series of big-picture lessons, with each chapter focusing on a single core lesson and related concepts. We will discuss a bit of theory when it will aid in understanding, but I will emphasize the practical application of the theory, often through specific, real-world examples. You will leave each chapter ready to apply the given lesson.

The lessons in the book are organized chronologically in the same way that I think about the storytelling with data process. Because of this and because later chapters do build on and in some cases refer back to earlier content, I recommend reading from beginning to end. After you've done this, you'll likely find yourself referring back to specific points of interest or examples that are relevant to the current data visualization challenges you face.

To give you a more specific idea of the path we'll take, chapter summaries can be found below.

Chapter 1: the importance of context

Before you start down the path of data visualization, there are a couple of questions that you should be able to concisely answer: Who is your audience? What do you need them to know or do? This chapter describes the importance of understanding the situational context, including the audience, communication mechanism, and desired tone. A number of concepts are introduced and illustrated via example to help ensure that context is fully understood. Creating a robust understanding of the situational context reduces iterations down the road and sets you on the path to success when it comes to creating visual content.

Chapter 2: choosing an effective visual

What is the best way to show the data you want to communicate? I've analyzed the visual displays I use most in my work. In this chapter, I introduce the most common types of visuals used to communicate data in a business setting, discuss appropriate use cases for each, and illustrate each through real-world examples. Specific types of visuals covered include simple text, table, heatmap, line graph, slopegraph, vertical bar chart, vertical stacked bar chart, waterfall chart, horizontal bar chart, horizontal stacked bar chart, and square area graph. We also cover visuals to be avoided, including pie and donut charts, and discuss reasons for avoiding 3D.

Chapter 3: clutter is your enemy!

Picture a blank page or a blank screen: every single element you add to that page or screen takes up cognitive load on the part of your audience. That means we should take a discerning eye to the elements we allow on our page or screen and work to identify those things that are taking up brain power unnecessarily and remove them. Identifying and eliminating clutter is the focus of this chapter. As part of this conversation, I introduce and discuss the Gestalt Principles of Visual Perception and how we can apply them to visual displays of information such as tables and graphs. We also discuss alignment, strategic use of white space, and contrast as important components of thoughtful design. Several examples are used to illustrate the lessons.

Chapter 4: focus your audience's attention

In this chapter, we continue to examine how people see and how you can use that to your advantage when crafting visuals. This includes a brief discussion on sight and memory that will act to frame up the importance of preattentive attributes like size, color, and position on page. We explore how preattentive attributes can be used strategically to help direct your audience's attention to where you want them to focus and to create a visual hierarchy of components to help direct your audience through the information you want to communicate in the way you want them to process it. Color as a strategic tool is covered in depth. Concepts are illustrated through a number of examples.

Chapter 5: think like a designer

Form follows function. This adage of product design has clear application to communicating with data. When it comes to the form and function of our data visualizations, we first want to think about what it is we want our audience to be able to *do* with the data (function) and create a visualization (form) that will allow for this with ease. In this chapter, we discuss how traditional design concepts can be applied to communicating with data. We explore affordances, accessibility, and aesthetics, drawing upon a number of concepts introduced previously, but looking at them through a slightly different lens. We also discuss strategies for gaining audience acceptance of your visual designs.

Chapter 6: dissecting model visuals

Much can be learned from a thorough examination of effective visual displays. In this chapter, we look at five exemplary visuals and discuss the specific thought process and design choices that led to their creation, utilizing the lessons covered up to this point. We explore decisions regarding the type of graph and ordering of data within the visual. We consider choices around what and how to emphasize and de-emphasize through use of color, thickness of lines, and relative size. We discuss alignment and positioning of components within the visuals and also the effective use of words to title, label, and annotate.

Chapter 7: lessons in storytelling

Stories resonate and stick with us in ways that data alone cannot. In this chapter, I introduce concepts of storytelling that can be leveraged for communicating with data. We consider what can be learned from master storytellers. A story has a clear beginning, middle, and end; we discuss how this framework applies to and can be used when constructing business presentations. We cover strategies for effective storytelling, including the power of repetition, narrative flow, considerations with spoken and written narratives, and various tactics to ensure that our story comes across clearly in our communications.

Chapter 8: pulling it all together

Previous chapters included piecemeal applications to demonstrate individual lessons covered. In this comprehensive chapter, we follow the storytelling with data process from start to finish using a single real-world example. We understand the context, choose an appropriate visual display, identify and eliminate clutter, draw attention to where we want our audience to focus, think like a designer, and tell a story. Together, these lessons and resulting visuals and narrative illustrate how we can move from simply showing data to telling a story with data.

Chapter 9: case studies

The penultimate chapter explores specific strategies for tackling common challenges faced in communicating with data through a number of case studies. Topics covered include color considerations with a dark background, leveraging animation in the visuals you present versus those you circulate, establishing logic in order, strategies for avoiding the spaghetti graph, and alternatives to pie charts.

Chapter 10: final thoughts

Data visualization—and communicating with data in general—sits at the intersection of science and art. There is certainly some science to it: best practices and guidelines to follow. There is also an artistic component. Apply the lessons we've covered to forge *your* path, using your artistic license to make the information easier for your audience to understand. In this final chapter, we discuss tips on where to go from here and strategies for upskilling storytelling with data competency in your team and your organization. We end with a recap of the main lessons covered.

Collectively, the lessons we'll cover will enable you to tell stories with data. Let's get started!

chapter one

the importance of context

This may sound counterintuitive, but success in data visualization does not start with data visualization. Rather, before you begin down the path of creating a data visualization or communication, attention and time should be paid to understanding the **context** for the need to communicate. In this chapter, we will focus on understanding the important components of context and discuss some strategies to help set you up for success when it comes to communicating visually with data.

Exploratory vs. explanatory analysis

Before we get into the specifics of context, there is one important distinction to draw, between *exploratory* and *explanatory* analysis. Exploratory analysis is what you do to understand the data and figure out what might be noteworthy or interesting to highlight to others. When we do exploratory analysis, it's like hunting for pearls in oysters.

We might have to open 100 oysters (test 100 different hypotheses or look at the data in 100 different ways) to find perhaps two pearls. When we're at the point of communicating our analysis to our audience, we really want to be in the *explanatory* space, meaning you have a specific thing you want to explain, a specific story you want to tell—probably about those two pearls.

Too often, people err and think it's OK to show exploratory analysis (simply present the data, all 100 oysters) when they should be showing explanatory (taking the time to turn the data into information that can be consumed by an audience: the two pearls). It is an understandable mistake. After undertaking an entire analysis, it can be tempting to want to show your audience *everything*, as evidence of all of the work you did and the robustness of the analysis. Resist this urge. You are making your audience reopen all of the oysters! Concentrate on the pearls, the information your audience needs to know.

Here, we focus on **explanatory** analysis and communication.

Recommended reading

For those interested in learning more about *exploratory* analysis, check out Nathan Yau's book, *Data Points*. Yau focuses on data visualization as a medium, rather than a tool, and spends a good portion of the book discussing the data itself and strategies for exploring and analyzing it.

Who, what, and how

When it comes to explanatory analysis, there are a few things to think about and be extremely clear on before visualizing any data or creating content. First, *To whom are you communicating?* It is important to have a good understanding of who your audience is and how they perceive you. This can help you to identify common ground that will help you ensure they hear your message. Second, *What do you want your audience to know or do?* You should be clear how you want your audience to act and take into account how you will communicate to them and the overall tone that you want to set for your communication.

It's only after you can concisely answer these first two questions that you're ready to move forward with the third: *How can you use data to help make your point?*

Let's look at the context of who, what, and how in a little more detail.

Who

Your audience

The more specific you can be about who your audience is, the better position you will be in for successful communication. Avoid general audiences, such as "internal and external stakeholders" or "anyone who might be interested"—by trying to communicate to too many different people with disparate needs at once, you put yourself in a position where you can't communicate to any one of them as effectively as you could if you narrowed your target audience. Sometimes this means creating different communications for different audiences. Identifying the decision maker is one way of narrowing your audience. The more you know about your audience, the better positioned you'll be to understand how to resonate with them and form a communication that will meet their needs and yours.

You

It's also helpful to think about the relationship that you have with your audience and how you expect that they will perceive you. Will you be encountering each other for the first time through this communication, or do you have an established relationship? Do they already trust you as an expert, or do you need to work to establish credibility? These are important considerations when it comes to determining how to structure your communication and whether and when to use data, and may impact the order and flow of the overall story you aim to tell.

Recommended reading

In Nancy Duarte's book *Resonate*, she recommends thinking of your audience as the hero and outlines specific strategies for getting to know your audience, segmenting your audience, and creating common ground. A free multimedia version of *Resonate* is available at duarte.com.

What

Action

What do you need your audience to know or do? This is the point where you think through how to make what you communicate relevant for your audience and form a clear understanding of why they should care about what you say. You should always want your audience to know or do something. If you can't concisely articulate that, you should revisit whether you need to communicate in the first place.

This can be an uncomfortable space for many. Often, this discomfort seems to be driven by the belief that the audience knows better than the presenter and therefore should choose whether and how to act on the information presented. This assumption is false. If you are the one analyzing and communicating the data, *you* likely know it best—*you* are a subject matter expert. This puts you in a unique position to interpret the data and help lead people to understanding and action. In general, those communicating with data need to take a more confident stance when it comes to making specific observations and recommendations based on their analysis. This will feel outside of your comfort zone if you haven't been routinely doing it.

Start doing it now—it will get easier with time. And know that even if you highlight or recommend the wrong thing, it prompts the right sort of conversation focused on action.

When it really isn't appropriate to recommend an action explicitly, encourage discussion toward one. Suggesting possible next steps can be a great way to get the conversation going because it gives your audience something to react to rather than starting with a blank slate. If you simply present data, it's easy for your audience to say, "Oh, that's interesting," and move on to the next thing. But if you ask for action, your audience has to make a decision whether to comply or not. This elicits a more productive reaction from your audience, which can lead to a more productive conversation—one that might never have been started if you hadn't recommended the action in the first place.

Prompting action

Here are some action words to help act as thought starters as you determine what you are asking of your audience:

accept | agree | begin | believe | change | collaborate | commence | create | defend | desire | differentiate | do | empathize | empower | encourage | engage | establish | examine | facilitate | familiarize | form | implement | include | influence | invest | invigorate | know | learn | like | persuade | plan | promote | pursue | recommend | receive | remember | report | respond | secure | support | simplify | start | try | understand | validate

Mechanism

How will you communicate to your audience? The method you will use to communicate to your audience has implications on a number of factors, including the amount of control you will have over how the audience takes in the information and the level of detail that needs to be explicit. We can think of the communication mechanism along a continuum, with live presentation at the left and a written document or email at the right, as shown in Figure 1.1. Consider the level of control you have over how the information is consumed as well as the amount of detail needed at either end of the spectrum.

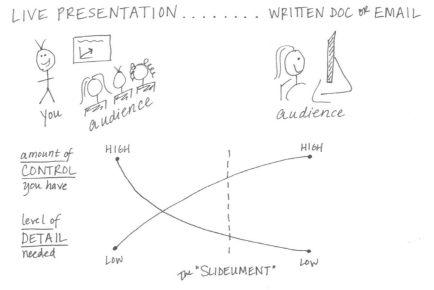

FIGURE 1.1   Communication mechanism continuum

At the left, with a **live presentation**, you (the presenter) are in full control. You determine what the audience sees and when they see it. You can respond to visual cues to speed up, slow down, or go into a particular point in more or less detail. Not all of the detail needs to be directly in the communication (the presentation or slide deck), because you, the subject matter expert, are there to answer any questions that arise over the course of the presentation and should be able and prepared to do so irrespective of whether that detail is in the presentation itself.

For live presentations, practice makes perfect

Do not use your slides as your teleprompter! If you find yourself reading each slide out loud during a presentation, you are using them as one. This creates a painful audience experience. You have to know your content to give a good presentation and this means practice, practice, and more practice! Keep your slides sparse, and only put things on them that help reinforce what you will say. Your slides can remind you of the next topic, but shouldn't act as your speaking notes.

Here are a few tips for getting comfortable with your material as you prepare for your presentation:

- Write out speaking notes with the important points you want to make with each slide.
- Practice what you want to say out loud to yourself: this ignites a different part of the brain to help you remember your talking points. It also forces you to articulate the transitions between slides that sometimes trip up presenters.
- Give a mock presentation to a friend or colleague.

At the right side of the spectrum, with a **written document or email**, you (the creator of the document or email) have less control. In this case, the audience is in control of how they consume the information. The level of detail that is needed here is typically higher because you aren't there to see and respond to your audience's cues. Rather, the document will need to directly address more of the potential questions.

In an ideal world, the work product for the two sides of this continuum would be totally different—sparse slides for a live presentation (since you're there to explain anything in more detail as needed), and denser documents when the audience is left to consume on their own. But in reality—due to time and other constraints—it is often the same product that is created to try to meet both of these needs. This gives rise to the **slideument**, a single document that's meant to solve both of these needs. This poses some challenges because of the diverse needs it is meant to satisfy, but we'll look at strategies for addressing and overcoming these challenges later in the book.

At this point at the onset of the communication process, it is important to identify the primary communication vehicle you'll be leveraging: live presentation, written document, or something else. Considerations on how much control you'll have over how your audience consumes the information and the level of detail needed will become very important once you start to generate content.

Tone

What tone do you want your communication to set? Another important consideration is the tone you want your communication to convey to your audience. Are you celebrating a success? Trying to light a fire to drive action? Is the topic lighthearted or serious? The tone you desire for your communication will have implications on the design choices that we will discuss in future chapters. For now, think about and specify the general tone that you want to establish when you set out on the data visualization path.

How

Finally—and only after we can clearly articulate who our audience is and what we need them to know or do—we can turn to the data and ask the question: *What data is available that will help make my point?* Data becomes supporting evidence of the story you will build and tell. We'll discuss much more on how to present this data visually in subsequent chapters.

Ignore the nonsupporting data?

You might assume that showing only the data that backs up your point and ignoring the rest will make for a stronger case. I do not recommend this. Beyond being misleading by painting a one-sided story, this is very risky. A discerning audience will poke holes in a story that doesn't hold up or data that shows one aspect but ignores the rest. The right amount of context and supporting and opposing data will vary depending on the situation, the level of trust you have with your audience, and other factors.

Who, what, and how: illustrated by example

Let's consider a specific example to illustrate these concepts. Imagine you are a fourth grade science teacher. You just wrapped up an experimental pilot summer learning program on science that was aimed at giving kids exposure to the unpopular subject. You surveyed the children at the onset and end of the program to understand whether and how perceptions toward science changed. You believe the data shows a great success story. You would like to continue to offer the summer learning program on science going forward.

Let's start with the *who* by identifying our audience. There are a number of different potential audiences who might be interested in this information: parents of students who participated in the program, parents of prospective future participants, the future potential participants themselves, other teachers who might be interested in doing something similar, or the budget committee that controls the funding you need to continue the program. You can imagine how the story you would tell to each of these audiences might differ. The emphasis might change. The call to action would be different for the different groups. The data you would show (or the decision to show data at all) could be different for the various audiences. You can imagine how, if we crafted a single communication meant to address all of these disparate audiences' needs, it would likely not exactly meet any single audience's need. This illustrates the importance of identifying a *specific* audience and crafting a communication with that specific audience in mind.

Let's assume in this case the audience we want to communicate to is the budget committee, which controls the funding we need to continue the program.

Now that we have answered the question of *who*, the *what* becomes easier to identify and articulate. If we're addressing the budget committee, a likely focus would be to demonstrate the success of the program and ask for a specific funding amount to continue to offer it. After identifying who our audience is and what we need from them, next we can think about the data we have available that will act as evidence of the story we want to tell. We can leverage the data collected via survey at the onset and end of the program to illustrate the increase in positive perceptions of science before and after the pilot summer learning program.

This won't be the last time we'll consider this example. Let's recap who we have identified as our audience, what we need them to know and do, and the data that will help us make our case:

> **Who:** The budget committee that can approve funding for continuation of the summer learning program.
>
> **What:** The summer learning program on science was a success; please approve budget of $X to continue.
>
> **How:** Illustrate success with data collected through the survey conducted before and after the pilot program.

Consulting for context: questions to ask

Often, the communication or deliverable you are creating is at the request of someone else: a client, a stakeholder, or your boss. This means you may not have all of the context and might need to consult with the requester to fully understand the situation. There is some-times additional context in the head of this requester that they may assume is known or not think to say out loud. Following are some questions you can use as you work to tease out this information. If you're on the requesting side of the communication and asking your support team to build a communication, think about answering these questions for them up front:

- What background information is relevant or essential?
- Who is the audience or decision maker? What do we know about them?
- What biases does our audience have that might make them sup-portive of or resistant to our message?
- What data is available that would strengthen our case? Is our audi-ence familiar with this data, or is it new?
- Where are the risks: what factors could weaken our case and do we need to proactively address them?
- What would a successful outcome look like?
- If you only had a limited amount of time or a single sentence to tell your audience what they need to know, what would you say?

In particular, I find that these last two questions can lead to insight-ful conversation. Knowing what the desired outcome is before you start preparing the communication is critical for structuring it well. Putting a significant constraint on the message (a short amount of time or a single sentence) can help you to boil the overall com-munication down to the single, most important message. To that end, there are a couple of concepts I recommend knowing and employing: the 3-minute story and the Big Idea.

The 3-minute story & Big Idea

The idea behind each of these concepts is that you are able to boil the "so-what" down to a paragraph and, ultimately, to a single, concise statement. You have to really know your stuff—know what the most important pieces are as well as what *isn't* essential in the most stripped-down version. While it sounds easy, being concise is often more challenging than being verbose. Mathematician and philosopher Blaise Pascal recognized this in his native French, with a statement that translates roughly to "I would have written a shorter letter, but I did not have the time" (a sentiment often attributed to Mark Twain).

3-minute story

The 3-minute story is exactly that: if you had only three minutes to tell your audience what they need to know, what would you say? This is a great way to ensure you are clear on and can articulate the story you want to tell. Being able to do this removes you from dependence on your slides or visuals for a presentation. This is useful in the situation where your boss asks you what you're working on or if you find yourself in an elevator with one of your stakeholders and want to give her the quick rundown. Or if your half-hour on the agenda gets shortened to ten minutes, or to five. If you know exactly what it is you want to communicate, you can make it fit the time slot you're given, even if it isn't the one for which you are prepared.

Big Idea

The Big Idea boils the so-what down even further: to a single sentence. This is a concept that Nancy Duarte discusses in her book, *Resonate* (2010). She says the Big Idea has three components:

1. It must articulate your unique point of view;

2. It must convey what's at stake; and

3. It must be a complete sentence.

Let's consider an illustrative 3-minute story and Big Idea, leveraging the summer learning program on science example that was introduced previously.

3-minute story: *A group of us in the science department were brainstorming about how to resolve an ongoing issue we have with incoming fourth-graders. It seems that when kids get to their first science class, they come in with this attitude that it's going to be difficult and they aren't going to like it. It takes a good amount of time at the beginning of the school year to get beyond that. So we thought, what if we try to give kids exposure to science sooner? Can we influence their perception? We piloted a learning program last summer aimed at doing just that. We invited elementary school students and ended up with a large group of second- and third-graders. Our goal was to give them earlier exposure to science in hopes of forming positive perception. To test whether we were successful, we surveyed the students before and after the program. We found that, going into the program, the biggest segment of students, 40%, felt just "OK" about science, whereas after the program, most of these shifted into positive perceptions, with nearly 70% of total students expressing some level of interest toward science. We feel that this demonstrates the success of the program and that we should not only continue to offer it, but also to expand our reach with it going forward.*

Big Idea: *The pilot summer learning program was successful at improving students' perceptions of science and, because of this success, we recommend continuing to offer it going forward; please approve our budget for this program.*

When you've articulated your story this clearly and concisely, creating content for your communication becomes much easier. Let's shift gears now and discuss a specific strategy when it comes to planning content: storyboarding.

Storyboarding

Storyboarding is perhaps the single most important thing you can do up front to ensure the communication you craft is on point. The storyboard establishes a structure for your communication. It is a visual outline of the content you plan to create. It can be subject to change as you work through the details, but establishing a structure early on will set you up for success. When you can (and as makes sense), get acceptance from your client or stakeholder at this step. It will help ensure that what you're planning is in line with the need.

When it comes to storyboarding, the biggest piece of advice I have is this: don't start with presentation software. It is too easy to go into slide-generating mode without thinking about how the pieces fit together and end up with a massive presentation deck that says nothing effectively. Additionally, as we start creating content via our computer, something happens that causes us to form an attachment to it. This attachment can be such that, even if we know what we've created isn't exactly on the mark or should be changed or eliminated, we are sometimes resistant to doing so because of the work we've already put in to get it to where it is.

Avoid this unnecessary attachment (and work!) by starting low tech. Use a whiteboard, Post-it notes, or plain paper. It's much easier to put a line through an idea on a piece of paper or recycle a Post-it note without feeling the same sense of loss as when you cut something you've spent time creating with your computer. I like using Post-it notes when I storyboard because you can rearrange (and add and remove) the pieces easily to explore different narrative flows.

If we storyboard our communication for the summer learning program on science, it might look something like Figure 1.2.

Note that in this example storyboard, the Big Idea is at the end, in the recommendation. Perhaps we'd want to consider leading with that to ensure that our audience doesn't miss the main point and to help set up why we are communicating to them and why they should care in the first place. We'll discuss additional considerations related to the narrative order and flow in Chapter 7.

FIGURE 1.2   Example storyboard

In closing

When it comes to explanatory analysis, being able to concisely articulate exactly who you want to communicate to and what you want
to convey before you start to build content reduces iterations and
helps ensure that the communication you build meets the intended
purpose. Understanding and employing concepts like the 3-minute
story, the Big Idea, and storyboarding will enable you to clearly and
succinctly tell your story and identify the desired flow.

While pausing before actually building the communication might feel
like it's a step that slows you down, in fact it helps ensure that you
have a solid understanding of what you want to do before you start
creating content, which will save you time down the road.

With that, consider your first lesson learned. You now **understand
the importance of context**.

chapter two

choosing an effective visual

There are many different graphs and other types of visual displays of information, but a handful will work for the majority of your needs. When I look back over the 150+ visuals that I created for workshops and consulting projects in the past year, there were only a dozen different types of visuals that I used (Figure 2.1). These are the visuals we'll focus on in this chapter.

91%

Simple text

Scatterplot

	A	B	C
Category 1	15%	22%	42%
Category 2	40%	36%	20%
Category 3	35%	17%	34%
Category 4	30%	29%	26%
Category 5	55%	30%	58%
Category 6	11%	25%	49%

Table

Line

	A	B	C
Category 1	15%	22%	42%
Category 2	40%	36%	20%
Category 3	35%	17%	34%
Category 4	30%	29%	26%
Category 5	55%	30%	58%
Category 6	11%	25%	49%

Heatmap

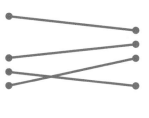

Slopegraph

FIGURE 2.1   The visuals I use most

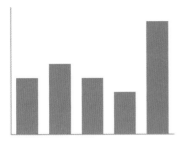

Vertical bar

Horizontal bar

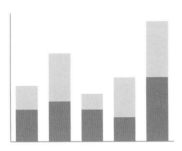

Stacked vertical bar

Stacked horizontal bar

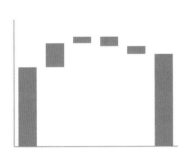

Waterfall

Square area

Simple text

When you have just a number or two to share, simple text can be a great way to communicate. Think about solely using the number—making it as prominent as possible—and a few supporting words to clearly make your point. Beyond potentially being misleading, putting one or only a couple of numbers in a table or graph simply causes the numbers to lose some of their oomph. When you have a number or two that you want to communicate, think about using the numbers themselves.

To illustrate this concept, let's consider the following example. A graph similar to Figure 2.2 accompanied an April 2014 Pew Research Center report on stay-at-home moms.

Children with a "Traditional" Stay-at-Home Mother

% of children with a married stay-at-home mother with a working husband

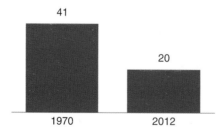

Note: Based on children younger than 18. Their mothers are categorized based on employment status in 1970 and 2012.

Source: Pew Research Center analysis of March Current Population Surveys Integrated Public Use Microdata Series (IPUMS-CPS), 1971 and 2013

Adapted from **PEW RESEARCH CENTER**

FIGURE 2.2   Stay-at-home moms original graph

The fact that you have some numbers does not mean that you need a graph! In Figure 2.2, quite a lot of text and space are used for a grand total of *two* numbers. The graph doesn't do much to aid in the interpretation of the numbers (and with the positioning of the data labels outside of the bars, it can even skew your perception of relative height such that 20 is less than half of 41 doesn't really come across visually).

In this case, a simple sentence would suffice: *20% of children had a traditional stay-at-home mom in 2012, compared to 41% in 1970.*

Alternatively, in a presentation or report, your visual could look something like Figure 2.3.

20%

of children had a
traditional stay-at-home mom
in 2012, compared to 41% in 1970

FIGURE 2.3   Stay-at-home moms simple text makeover

As a side note, one consideration in this specific example might be whether you want to show an entirely different metric. For example, you could reframe in terms of the percent change: "The number of children having a traditional stay-at-home mom decreased more than 50% between 1970 and 2012." I advise caution, however, any time you reduce from multiple numbers down to a single one—think about what context may be lost in doing so. In this case, I find that the actual magnitude of the numbers (20% and 41%) is helpful in interpreting and understanding the change.

When you have just a number or two that you want to communicate: *use the numbers directly.*

When you have more data that you want to show, generally a table or graph is the way to go. One thing to understand is that people interact differently with these two types of visuals. Let's discuss each in detail and look at some specific varieties and use cases.

Tables

Tables interact with our verbal system, which means that we *read* them. When I have a table in front of me, I typically have my index finger out: I'm reading across rows and down columns or I'm comparing values. Tables are great for just that—communicating to a mixed audience whose members will each look for their particular row of interest. If you need to communicate multiple different units of measure, this is also typically easier with a table than a graph.

Tables in live presentations

Using a table in a live presentation is rarely a good idea. As your audience reads it, you lose their ears and attention to make your point verbally. When you find yourself using a table in a presentation or report, ask yourself: what is the point you are trying to make? Odds are that there will be a better way to pull out and visualize the piece or pieces of interest. In the event that you feel you're losing too much by doing this, consider whether including the full table in the appendix and a link or reference to it will meet your audience's needs.

One thing to keep in mind with a table is that you want the design to fade into the background, letting the data take center stage. Don't let heavy borders or shading compete for attention. Instead, think of using light borders or simply white space to set apart elements of the table.

Take a look at the example tables in Figure 2.4. As you do, note how the data stands out more than the structural components of the table in the second and third iterations (light borders, minimal borders).

Heavy borders

Group	Metric A	Metric B	Metric C
Group 1	$X.X	Y%	Z,ZZZ
Group 2	$X.X	Y%	Z,ZZZ
Group 3	$X.X	Y%	Z,ZZZ
Group 4	$X.X	Y%	Z,ZZZ
Group 5	$X.X	Y%	Z,ZZZ

Light borders

Group	Metric A	Metric B	Metric C
Group 1	$X.X	Y%	Z,ZZZ
Group 2	$X.X	Y%	Z,ZZZ
Group 3	$X.X	Y%	Z,ZZZ
Group 4	$X.X	Y%	Z,ZZZ
Group 5	$X.X	Y%	Z,ZZZ

Minimal borders

Group	Metric A	Metric B	Metric C
Group 1	$X.X	Y%	Z,ZZZ
Group 2	$X.X	Y%	Z,ZZZ
Group 3	$X.X	Y%	Z,ZZZ
Group 4	$X.X	Y%	Z,ZZZ
Group 5	$X.X	Y%	Z,ZZZ

FIGURE 2.4   Table borders

Borders should be used to improve the legibility of your table. Think about pushing them to the background by making them grey, or getting rid of them altogether. The data should be what stands out, not the borders.

Recommended reading

For more on table design, check out Stephen Few's book, *Show Me the Numbers*. There is an entire chapter dedicated to the design of tables, with discussion on the structural components of tables and best practices in table design.

Next, let's shift our focus to a special case of tables: the heatmap.

Heatmap

One approach for mixing the detail you can include in a table while also making use of visual cues is via a heatmap. A heatmap is a way to visualize data in tabular format, where in place of (or in addition to) the numbers, you leverage colored cells that convey the relative magnitude of the numbers.

Consider Figure 2.5, which shows some generic data in a table and also a heatmap.

Table

	A	B	C
Category 1	15%	22%	42%
Category 2	40%	36%	20%
Category 3	35%	17%	34%
Category 4	30%	29%	26%
Category 5	55%	30%	58%
Category 6	11%	25%	49%

Heatmap

LOW·HIGH

	A	B	C
Category 1	15%	22%	42%
Category 2	40%	36%	20%
Category 3	35%	17%	34%
Category 4	30%	29%	26%
Category 5	55%	30%	58%
Category 6	11%	25%	49%

FIGURE 2.5   Two views of the same data

In the table in Figure 2.5, you are left to read the data. I find myself scanning across rows and down columns to get a sense of what I'm looking at, where numbers are higher or lower, and mentally stack rank the categories presented in the table.

To reduce this mental processing, we can use **color saturation** to provide visual cues, helping our eyes and brains more quickly target the potential points of interest. In the second iteration of the table on the right entitled "Heatmap," the higher saturation of blue, the higher the number. This makes the process of picking out the tails of the spectrum—the lowest number (11%) and highest number (58%)—an easier and faster process than it was in the original table where we didn't have any visual cues to help direct our attention.

Graphing applications (like Excel) typically have conditional formatting functionality built in that allows you to apply formatting like that shown in Figure 2.5 with ease. Be sure when you leverage this to always include a legend to help the reader interpret the data (in this case, the LOW-HIGH subtitle on the heatmap with color corresponding to the conditional formatting color serves this purpose).

Next, let's shift our discussion to the visuals we tend to think of first when it comes to communicating with data: graphs.

Graphs

While tables interact with our verbal system, graphs interact with our visual system, which is faster at processing information. This means that a well-designed graph will typically get the information across more quickly than a well-designed table. As I mentioned at the onset of this chapter, there are a plethora of graph types out there. The good news is that a handful of them will meet most of your everyday needs.

The types of graphs I frequently use fall into four categories: points, lines, bars, and area. We will examine these more closely and discuss the subtypes that I find myself using on a regular basis, with specific use cases and examples for each.

Chart or graph?

Some draw a distinction between charts and graphs. Typically, "chart" is the broader category, with "graphs" being one of the subtypes (other chart types include maps and diagrams). I don't tend to draw this distinction, since nearly all of the charts I deal with on a regular basis are graphs. Throughout this book, I use the words *chart* and *graph* interchangeably.

Points

Scatterplot

Scatterplots can be useful for showing the relationship between two things, because they allow you to encode data simultaneously on a horizontal x-axis and vertical y-axis to see whether and what relationship exists. They tend to be more frequently used in scientific fields (and perhaps, because of this, are sometimes viewed as complicated to understand by those less familiar with them). Though infrequent, there are use cases for scatterplots in the business world as well.

For example, let's say that we manage a bus fleet and want to understand the relationship between miles driven and cost per mile. The scatterplot may look something like Figure 2.6.

Cost per mile by miles driven

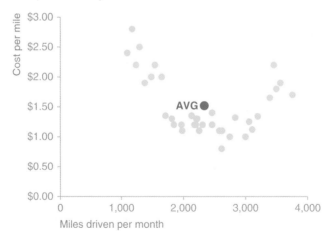

FIGURE 2.6 Scatterplot

If we want to focus primarily on those cases where cost per mile is above average, a slightly modified scatterplot designed to draw our eye there more quickly might look something like what is shown in Figure 2.7.

Cost per mile by miles driven

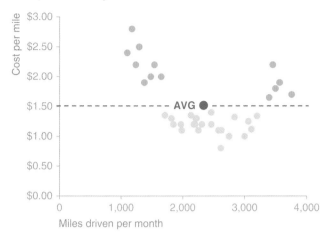

FIGURE 2.7   Modified scatterplot

We can use Figure 2.7 to make observations such as cost per mile is higher than average when less than about 1,700 miles or more than about 3,300 miles were driven for the sample observed. We'll talk more about the design choices made here and reasons for them in upcoming chapters.

Lines

Line graphs are most commonly used to plot continuous data. Because the points are physically connected via the line, it implies a connection between the points that may not make sense for categorical data (a set of data that is sorted or divided into different categories). Often, our continuous data is in some unit of time: days, months, quarters, or years.

Within the line graph category, there are two types of charts that I frequently find myself using: the standard line graph and the slopegraph.

Line graph

The line graph can show a single series of data, two series of data, or multiple series, as illustrated in Figure 2.8.

FIGURE 2.8   Line graphs

Note that when you're graphing time on the horizontal x-axis of a line graph, the data plotted must be in consistent intervals. I recently saw a graph where the units on the x-axis were decades from 1900 forward (1910, 1920, 1930, etc.) and then switched to yearly after 2010 (2011, 2012, 2013, 2014). This meant that the distance between the decade points and annual points looked the same. This is a misleading way to show the data. Be consistent in the time points you plot.

Showing average within a range in a line graph

In some cases, the line in your line graph may represent a summary statistic, like the average, or the point estimate of a forecast. If you also want to give a sense of the range (or confidence level, depending on the situation), you can do that directly on the graph by also visualizing this range. For example, the graph in Figure 2.9 shows the minimum, average, and maximum wait times at passport control for an airport over a 13-month period.

Passport control wait time
Past 13 months

FIGURE 2.9   Showing average within a range in a line graph

Slopegraph

Slopegraphs can be useful when you have two time periods or points of comparison and want to quickly show relative increases and decreases or differences across various categories between the two data points.

The best way to explain the value of and use case for slopegraphs is through a specific example. Imagine that you are analyzing and communicating data from a recent employee feedback survey. To show the relative change in survey categories from 2014 to 2015, the slopegraph might look something like Figure 2.10.

Slopegraphs pack in a lot of information. In addition to the absolute values (the points), the lines that connect them give you the visual increase or decrease in rate of change (via the slope or direction) without ever having to explain that's what they are doing, or what exactly a "rate of change" is—rather, it's intuitive.

Employee feedback over time

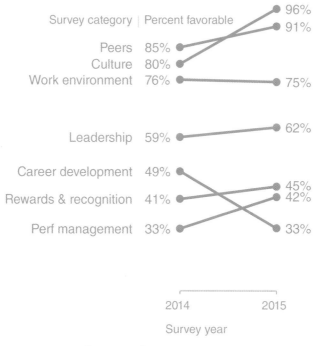

FIGURE 2.10   Slopegraph

Slopegraph template

S lopegraphs can take a bit of patience to set up because they often aren't one of the standard graphs included in graphing applications. An Excel template with an example slopegraph and instructions for customized use can be downloaded here: storytellingwithdata.com/slopegraph-template.

Whether a slopegraph will work in your specific situation depends on the data itself. If many of the lines are overlapping, a slopegraph may not work, though in some cases you can still emphasize a single series at a time with success. For example, we can draw attention to the single category that decreased over time from the preceding example.

Employee feedback over time

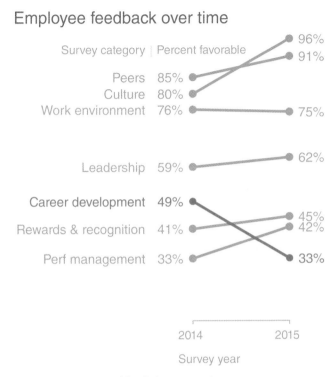

FIGURE 2.11   Modified slopegraph

In Figure 2.11, our attention is drawn immediately to the decrease in "Career development," while the rest of the data is preserved for context without competing for attention. We will talk about the strategy behind this when we discuss preattentive attributes in Chapter 4.

While lines work well to show data over time, bars tend to be my go-to graph type for plotting categorical data, where information is organized into groups.

Bars

Sometimes bar charts are avoided because they are common. This is a mistake. Rather, bar charts should be leveraged *because they are common,* as this means less of a learning curve for your audience. Instead of using their brain power to try to understand how to read the graph, your audience spends it figuring out what information to take away from the visual.

Bar charts are easy for our eyes to read. Our eyes compare the end points of the bars, so it is easy to see quickly which category is the biggest, which is the smallest, and also the incremental difference between categories. Note that, because of how our eyes compare the relative end points of the bars, it is important that bar charts always have a zero baseline (where the x-axis crosses the y-axis at zero), otherwise you get a false visual comparison.

Consider Figure 2.12 from Fox News.

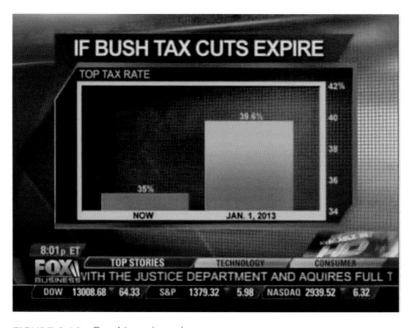

FIGURE 2.12   Fox News bar chart

For this example, let's imagine we are back in the fall of 2012. We are wondering what will happen if the Bush tax cuts expire. On the left-hand side, we have what the top tax rate is currently, 35%, and on the right-hand side what it will be as of January 1, at 39.6%.

When you look at this graph, how does it make you feel about the potential expiration of the tax cuts? Perhaps worried about the huge increase? Let's take a closer look.

Note that the bottom number on the vertical axis (shown at the far right) is not zero, but rather 34. This means that the bars, in theory, should continue down through the bottom of the page. In fact, the way this is graphed, the visual increase is *460%* (the heights of the bars are 35 – 34 = 1 and 39.6 – 34 = 5.6, so (5.6 – 1) / 1 = 460%). If we graph the bars with a zero baseline so that the heights are accurately represented (35 and 39.6), we get an actual visual increase of *13%* ((39.6 – 35) / 35). Let's look at a side-by-side comparison in Figure 2.13.

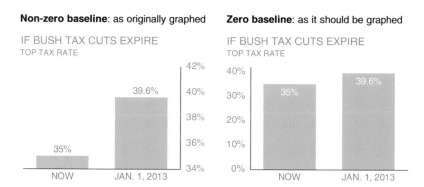

FIGURE 2.13   Bar charts must have a zero baseline

In Figure 2.13, what looked like a huge increase on the left is reduced considerably when plotted appropriately. Perhaps the tax increase isn't so worrisome, or at least not as severe as originally depicted. Because of the way our eyes compare the relative end points of the bars, it's important to have the context of the entire bar there in order to make an accurate comparison.

You'll note that a couple of other design changes were made in the remake of this visual as well. The y-axis labels that were placed on the right-hand side of the original visual were moved to the left (so we see how to interpret the data before we get to the actual data). The data labels that were originally outside of the bars were pulled inside to reduce clutter. If I were plotting this data outside of this specific lesson, I might omit the y-axis entirely and show only the data labels within the bars to reduce redundant information. However, in this case, I preserved the axis to make it clear that it begins at zero.

Graph axis vs. data labels

When graphing data, a common decision to make is whether to preserve the axis labels or eliminate the axis and instead label the data points directly. In making this decision, consider the level of specificity needed. If you want your audience to focus on big-picture trends, think about preserving the axis but deemphasizing it by making it grey. If the specific numerical values are important, it may be better to label the data points directly. In this latter case, it's usually best to omit the axis to avoid the inclusion of redundant information. Always consider how you want your audience to use the visual and construct it accordingly.

The rule we've illustrated here is that *bar charts must have a zero baseline*. Note that this rule does not apply to line graphs. With line graphs, since the focus is on the relative position in space (rather than the length from the baseline or axis), you can get away with a nonzero baseline. Still, you should approach with caution—make it clear to your audience that you are using a nonzero baseline and take context into account so you don't overzoom and make minor changes or differences appear significant.

Ethics and data visualization

But what if changing the scale on a bar chart or otherwise manipulating the data better reinforces the point you want to make? Misleading in this manner by inaccurately visualizing data is not OK. Beyond ethical concerns, it is risky territory. All it takes is one discerning audience member to notice the issue (for example, the y-axis of a bar chart beginning at something other than zero) and your entire argument will be thrown out the window, along with your credibility.

While we're considering lengths of bars, let's also spend a moment on the *width* of bars. There's no hard-and-fast rule here, but in general the bars should be wider than the white space between the bars. You don't want the bars to be so wide, however, that your audience wants to compare areas instead of lengths. Consider the following "Goldilocks" of bar charts: too thin, too thick, and just right.

FIGURE 2.14   Bar width

We've discussed some best practices when it comes to bar charts in general. Next let's take a look at some different varieties. Having a number of bar charts at your disposal gives you flexibility when facing different data visualization challenges. We'll look at the ones I think you should be familiar with here.

Vertical bar chart

The plain vanilla bar chart is the vertical bar chart, or column chart. Like line graphs, vertical bar charts can be single series, two series, or multiple series. Note that as you add more series of data, it becomes more difficult to focus on one at a time and pull out insight, so use multiple series bar charts with caution. Be aware also that there is visual grouping that happens as a result of the spacing in bar charts having more than one data series. This makes the relative order of the categorization important. Consider what you want your audience to be able to compare, and structure your categorization hierarchy to make that as easy as possible.

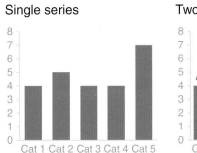
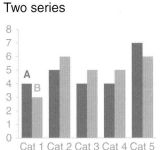
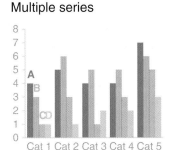

FIGURE 2.15 Bar charts

Stacked vertical bar chart

Use cases for stacked vertical bar charts are more limited. They are meant to allow you to compare totals across categories and also see the subcomponent pieces within a given category. This can quickly become visually overwhelming, however—especially given the varied default color schemes in most graphing applications (more to come on that). It is hard to compare the subcomponents across the various categories once you get beyond the bottom series (the one directly next to the x-axis) because you no longer have a consistent baseline to use to compare. This makes it a harder comparison for our eyes to make, as illustrated in Figure 2.16.

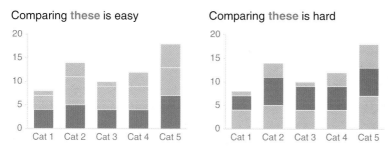

FIGURE 2.16   Comparing series with stacked bar charts

The stacked vertical bar chart can be structured as absolute numbers (where you plot the numbers directly, as shown in Figure 2.16), or with each column summing to 100% (where you plot the percent of total for each vertical segment; we'll look at a specific example of this in Chapter 9). Which you choose depends on what you are trying to communicate to your audience. When you use the 100% stacked bar, think about whether it makes sense to also include the absolute numbers for each category total (either in an unobtrusive way in the graph directly, or possibly in a footnote), which may aid in the interpretation of the data.

Waterfall chart

The waterfall chart can be used to pull apart the pieces of a stacked bar chart to focus on one at a time, or to show a starting point, increases and decreases, and the resulting ending point.

The best way to illustrate the use case for a waterfall chart is through a specific example. Imagine that you are an HR business partner and want to understand and communicate how employee headcount has changed over the past year for the client group you support.

A waterfall chart showing this breakdown might look something like Figure 2.17.

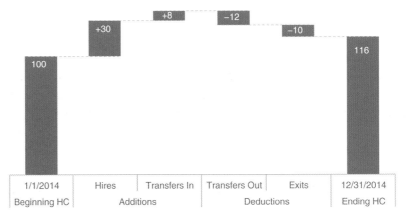

2014 Headcount math
Though more employees transferred out of the team than transferred in,
aggressive hiring means overall headcount (HC) increased 16% over the course of the year.

FIGURE 2.17   Waterfall chart

On the left-hand side, we see what the employee headcount for the given team was at the beginning of the year. As we move to the right, first we encounter the incremental additions: new hires and employees transferring into the team from other parts of the organization. This is followed by the deductions: transfers out of the team to other parts of the organization and attrition. The final column represents employee headcount at the end of the year, after the additions and deductions have been applied to the beginning of year headcount.

Brute-force waterfall charts

If your graphing application doesn't have waterfall chart functionality built in, fret not. The secret is to leverage the stacked bar chart and make the first series (the one that appears closest to the x-axis) invisible. It takes a bit of math to set up correctly, but it works great. A blog post on this topic, along with an example Excel version of the above chart and instructions on how to set one up for your own purposes can be downloaded at storytellingwithdata.com/ waterfall-chart.

Horizontal bar chart

If I had to pick a single go-to graph for categorical data, it would be the horizontal bar chart, which flips the vertical version on its side. Why? Because it is *extremely easy to read*. The horizontal bar chart is especially useful if your category names are long, as the text is written from left to right, as most audiences read, making your graph legible for your audience. Also, because of the way we typically process information—starting at top left and making z's with our eyes across the screen or page—the structure of the horizontal bar chart is such that our eyes hit the category names before the actual data. This means by the time we get to the data, we already know what it represents (instead of the darting back and forth our eyes do between the data and category names with vertical bar charts).

Like the vertical bar chart, the horizontal bar chart can be single series, two series, or multiple series (Figure 2.18).

FIGURE 2.18   Horizontal bar charts

The logical ordering of categories

When designing any graph showing categorical data, be thoughtful about how your categories are ordered. If there is a natural ordering to your categories, it may make sense to leverage that. For example, if your categories are age groups—0–10 years old, 11–20 years old, and so on—keep the categories in numerical order. If, however, there isn't a natural ordering in your categories that makes sense to leverage, think about what ordering of your data will make the most sense. Being thoughtful here can mean providing a construct for your audience, easing the interpretation process.

Your audience (without other visual cues) will typically look at your visual starting at the top left and zigzagging in "z" shapes. This means they will encounter the top of your graph first. If the biggest category is the most important, think about putting that first and ordering the rest of the categories in decreasing numerical order. Or if the smallest is most important, put that at the top and order by ascending data values.

For a specific example about the logical ordering of data, check out case study 3 in Chapter 9.

Stacked horizontal bar chart

Similar to the stacked vertical bar chart, stacked horizontal bar charts can be used to show the totals across different categories but also give a sense of the subcomponent pieces. They can be structured to show either absolute values or sum to 100%.

I find this latter approach can work well for visualizing portions of a whole on a scale from negative to positive, because you get a consistent baseline on both the far left and the far right, allowing for easy comparison of the left-most pieces as well as the right-most pieces. For example, this approach can work well for visualizing survey data collected along a Likert scale (a scale commonly used in surveys that typically ranges from Strongly Disagree to Strongly Agree), as shown in Figure 2.19.

Survey results

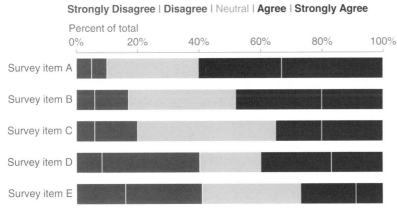

FIGURE 2.19   100% stacked horizontal bar chart

Area

I avoid most area graphs. Humans' eyes don't do a great job of attributing quantitative value to two-dimensional space, which can render area graphs harder to read than some of the other types of visual displays we've discussed. For this reason, I typically avoid them, with one exception—when I need to visualize numbers of vastly different magnitudes. The second dimension you get using a square for this (which has both height and width, compared to a bar that has only height *or* width) allows this to be done in a more compact way than possible with a single dimension, as shown in Figure 2.20.

Interview breakdown

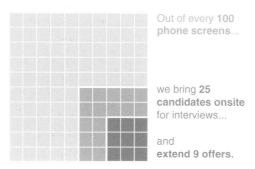

Out of every **100 phone screens**...

we bring **25 candidates onsite** for interviews...

and **extend 9 offers.**

FIGURE 2.20 Square area graph

Other types of graphs

What I've covered up to this point are the types of graphs I find myself commonly using. This is certainly not an exhaustive list. However, they should meet the majority of your everyday needs. Mastering the basics is imperative before exploring novel types of data visualization.

There are many other types of graphs out there. When it comes to selecting a graph, first and foremost, choose a graph type that will enable you to clearly get your message across to your audience. With less familiar types of visuals, you will likely need to take extra care in making them accessible and understandable.

Infographics

Infographic is a term that is frequently misused. An infographic is simply a graphical representation of information or data. Visuals coined *infographic* run the gamut from fluffy to informative. On the inadequate end of the spectrum, they often include elements like garish, oversized numbers and cartoonish graphics. These designs have a certain visual appeal and can seduce the reader. On second glance, however, they appear shallow and leave a discerning audience dissatisfied. Here, the description of "information graphic"—though often used—is not appropriate. On the other end of the spectrum are infographics that live up to their name and actually inform. There are many good examples in the area of data journalism (for example, the *New York Times* and *National Geographic*).

There are critical questions information designers must be able to answer before they begin the design process. These are the same questions we've discussed when it comes to understanding the context for storytelling with data. Who is your audience? What do you need them to know or do? It is only after the answers to these questions can be succinctly articulated that an effective method of display that will best aid the message can be chosen. Good data visualization—infographic or otherwise—is not simply a collection of facts on a given topic; good data visualization tells a story.

To be avoided

We've discussed the visuals that I use most commonly to communicate data in a business setting. There are also some specific graph types and elements that you should avoid: pie charts, donut charts, 3D, and secondary y-axes. Let's discuss each of these.

Pie charts are evil

I have a well-documented disdain for pie charts. In short, they are evil. To understand how I arrived at this conclusion, let's look at an example.

The pie chart shown in Figure 2.21 (based on a real example) shows market share across four suppliers: A, B, C, and D. If I asked you to make a simple observation—which supplier is the largest based on this visual—what would you say?

Supplier Market Share

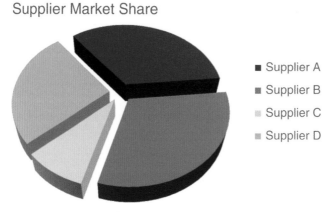

FIGURE 2.21   Pie chart

Most people will agree that "Supplier B," rendered in medium blue at the bottom right, appears to be the largest. If you had to estimate what proportion supplier B makes up of the overall market, what percent might you estimate?

35%?

40%?

Perhaps you can tell by my leading questioning that something fishy is going on here. Take a look at what happens when we add the numbers to the pie segments, as shown in Figure 2.22.

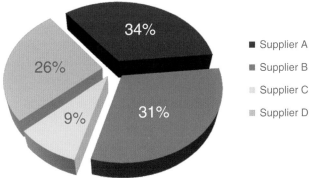

FIGURE 2.22   Pie chart with labeled segments

"Supplier B"—which *looks* largest, at 31%—is actually smaller than "Supplier A" above it, which looks smaller.

Let's discuss a couple of issues that pose a challenge for accurately interpreting this data. The first thing that catches your eye (and suspicion, if you're a discerning chart reader) is the 3D and strange perspective that's been applied to the graph, tilting the pie and making the pieces at the top appear farther away and thus smaller than they actually are, while the pieces at the bottom appear closer and thus bigger than they actually are. We'll talk more about 3D soon, but for now I'll articulate a relevant data visualization rule: *don't use 3D!* It does nothing good, and can actually do a whole lot of harm, as we see here with the way it skews the visual perception of the numbers.

Even when we strip away the 3D and flatten the pie, interpretation challenges remain. The human eye isn't good at ascribing quantitative value to two-dimensional space. Said more simply: *pie charts are hard for people to read.* When segments are close in size, it's difficult (if not impossible) to tell which is bigger. When they aren't close in size, the best you can do is determine that one is bigger than the other, but you can't judge by how much. To get over this, you can add data labels as has been done here. But I'd still argue the visual isn't worth the space it takes up.

What should you do instead? One approach is to replace the pie chart with a horizontal bar chart, as illustrated in Figure 2.23, organized from greatest to least or vice versa (unless there is some natural ordering to the categories that makes sense to leverage, as mentioned earlier). Remember, with bar charts, our eyes compare the end points. Because they are aligned at a common baseline, it is easy to assess relative size. This makes it straightforward to see not only which segment is the largest, for example, but also *how incrementally larger* it is than the other segments.

Supplier Market Share

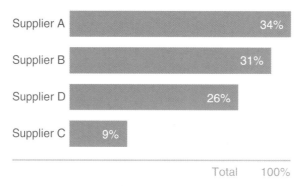

FIGURE 2.23   An alternative to the pie chart

One might argue that you lose something in the transition from pie to bar. The unique thing you get with a pie chart is the concept of there being a whole and, thus, parts of a whole. But if the visual is difficult to read, is it worth it? In Figure 2.23, I've tried to address this by showing that the pieces sum to 100%. It isn't a perfect solution, but something to consider. For more alternatives to pie charts, check out case study 5 in Chapter 9.

If you find yourself using a pie chart, pause and ask yourself: *why?* If you're able to answer this question, you've probably put enough thought into it to use the pie chart, but it certainly shouldn't be the first type of graph that you reach for, given some of the difficulties in visual interpretation we've discussed here.

While we're on the topic of pie charts, let's look quickly at another "dessert visual" to avoid: the donut chart.

The donut chart

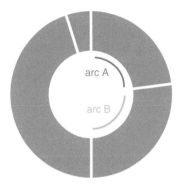

FIGURE 2.24   Donut chart

With pies, we are asking our audience to compare angles and areas. With a donut chart, we are asking our audience to compare one arc length to another arc length (for example, in Figure 2.24, the length of *arc A* compared to *arc B*). How confident do you feel in your eyes' ability to ascribe quantitative value to an arc length?

Not very? That's what I thought. Don't use donut charts.

Never use 3D

One of the golden rules of data visualization goes like this: never use 3D. Repeat after me: never use 3D. The only exception is if you are actually *plotting a third dimension* (and even then, things get really tricky really quickly, so take care when doing this)—and you should never use 3D to plot a single dimension. As we saw in the pie chart example previously, 3D skews our numbers, making them difficult or impossible to interpret or compare.

Adding 3D to graphs introduces unnecessary chart elements like side and floor panels. Even worse than these distractions, graphing applications do some pretty strange things when it comes to plotting values in 3D. For example, in a 3D bar chart, you might think that your graphing application plots the front of the bar or perhaps the back of the bar. Unfortunately, it's often even less straightforward than that. In Excel, for example, the bar height is determined by an invisible tangent plane intersecting the corresponding height on the y-axis. This gives rise to graphs like the one shown in Figure 2.25.

Number of issues

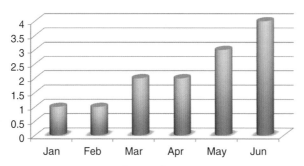

FIGURE 2.25   3D column chart

Judging by Figure 2.25, how many issues were there in January and February? I've plotted a single issue for each of these months. However, the way I read the chart, if I compare the bar height to the gridlines and follow it leftward to the y-axis, I'd estimate visually a value of maybe 0.8. This is simply bad data visualization. Don't use 3D.

Secondary y-axis: generally not a good idea

Sometimes it's useful to be able to plot data that is in entirely different units against the same x-axis. This often gives rise to the secondary y-axis: another vertical axis on the right-hand side of the graph. Consider the example shown in Figure 2.26.

Secondary y-axis

FIGURE 2.26   Secondary y-axis

When interpreting Figure 2.26, it takes some time and reading to understand which data should be read against which axis. Because of this, you should avoid the use of a secondary or right-hand y-axis. Instead, think about whether one of the following approaches will meet your needs:

1. Don't show the second y-axis. Instead, label the data points that belong on this axis directly.

2. Pull the graphs apart vertically and have a separate y-axis for each (both along the left) but leverage the same x-axis across both.

Figure 2.27 illustrates these options.

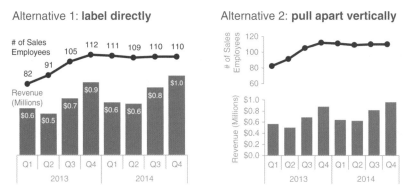

FIGURE 2.27   Strategies for avoiding a secondary y-axis

A third potential option not shown here is to link the axis to the data to be read against it through the use of color. For example, in the original graph depicted in Figure 2.26, I could write the left y-axis title "Revenue" in blue and keep the revenue bars blue while at the same time writing the right y-axis title "# of Sales Employees" in orange and making the line graph orange to tie these together visually. I don't recommend this approach because color can typically be used more strategically. We'll spend a lot more time discussing color in Chapter 4.

It is also worth noting that when you display two datasets against the same axis, it can imply a relationship that may or may not exist. This is something to be aware of when determining whether this is an appropriate approach in the first place.

When you're facing a secondary y-axis challenge and considering which alternative shown in Figure 2.27 will better meet your needs, think about the level of specificity you need. Alternative 1, where each data point is labeled explicitly, puts more attention on the specific numbers. Alternative 2, where the axes are shown at the left, puts more focus on the overarching trends. In general, avoid a secondary y-axis and instead employ one of these alternate approaches.

In closing

In this chapter, we've explored the types of visual displays I find myself using most. There will be use cases for other types of visuals, but what we've covered here should meet the majority of everyday needs.

In many cases, there isn't a single correct visual display; rather, often there are different types of visuals that could meet a given need. Drawing from the previous chapter on context, most important is to have that need clearly articulated: *What do you need your audience to know?* Then choose a visual display that will enable you to make this clear.

If you're wondering *What is the right graph for my situation?*, the answer is always the same: whatever will be easiest for your audience to read. There is an easy way to test this, which is to create your visual and show it to a friend or colleague. Have them articulate the following as they process the information: where they focus, what they see, what observations they make, what questions they have. This will help you assess whether your visual is hitting the mark, or in the case where it isn't, help you know where to concentrate your changes.

You now know the second lesson of storytelling with data: how to **choose an appropriate visual display**.

chapter three

clutter is your enemy!

Picture a blank page or a blank screen: every single element you add to that page or screen takes up cognitive load on the part of your audience—in other words, takes them brain power to process. Therefore, we want to take a discerning look at the visual elements that we allow into our communications. In general, identify anything that isn't adding informative value—or isn't adding *enough* informative value to make up for its presence—and remove those things. Identifying and eliminating such clutter is the focus of this chapter.

Cognitive load

You have felt the burden of cognitive load before. Perhaps you were sitting in a conference room as the person leading the meeting was flipping through their projected slides and they paused on one that looked overwhelmingly busy and complicated. Yikes, did you say "ugh" out loud, or was that just in your head? Or maybe you were reading through a report or the newspaper, and a graph caught your eye just long enough for you to think, "this looks interesting but I have no idea what I'm meant to get out of it"—and rather than spend more time to decipher it, you turned the page.

In both of these instances, what you've experienced is excessive or extraneous cognitive load.

We experience cognitive load *anytime* we take in information. Cognitive load can be thought of as the mental effort that's required to learn new information. When we ask a computer to do work, we are relying on the computer's processing power. When we ask our audience to do work, we are leveraging their mental processing power. This is cognitive load. Humans' brains have a finite amount of this mental processing power. As designers of information, we want to be smart about how we use our audience's brain power. The preceding examples point to extraneous cognitive load: processing that takes up mental resources but doesn't help the audience understand the information. This is something we want to avoid.

The data-ink or signal-to-noise ratio

A number of concepts have been introduced over time in an effort to explain and help provide guidance for reducing the cognitive load we push to our audience through our visual communications. In his book *The Visual Display of Quantitative Information*, Edward Tufte refers to maximizing the data-ink ratio, saying "the larger the share of a graphic's ink devoted to data, the better (other relevant matters being equal)." This can also be referred to as maximizing the signal-to-noise ratio (see Nancy Duarte's book *Resonate*), where the signal is the information we want to communicate, and the noise are those elements that either don't add to, or in some cases detract from, the message we are trying to impart to our audience.

What matters most when it comes to our visual communications is the *perceived* cognitive load on the part of our audience: how hard they believe they are going to have to work to get the information out of your communication. This is a decision they likely reach without giving it much (if any) conscious thought, and yet it can make the difference between getting your message across or not.

In general, think about minimizing the perceived cognitive load (to the extent that is reasonable and still allows you to get the information across) for your audience.

Clutter

One culprit that can contribute to excessive or extraneous cognitive load is something I refer to simply as **clutter**. These are visual elements that take up space but don't increase understanding. We'll take a more specific look at exactly what elements can be considered clutter soon, but in the meantime I want to talk generally about why clutter is a bad thing.

There is a simple reason we should aim to reduce clutter: because it makes our visuals appear more complicated than necessary.

Perhaps without explicitly recognizing it, the presence of clutter in our visual communications can cause a less-than-ideal—or worse—uncomfortable user experience for our audience (this is that "ugh" moment I referred to at the beginning of this chapter). Clutter can make something feel more complicated than it actually is. When our visuals feel complicated, we run the risk of our audience deciding they don't want to take the time to understand what we're showing, at which point we've lost our ability to communicate with them. This is not a good thing.

Gestalt principles of visual perception

When it comes to identifying which elements in our visuals are signal (the information we want to communicate) and which might be noise (clutter), consider the **Gestalt Principles of Visual Perception**. The Gestalt School of Psychology set out in the early 1900s to understand how individuals perceive order in the world around them. What they came away with are the principles of visual perception still accepted today that define how people interact with and create order out of visual stimuli.

We'll discuss six principles here: proximity, similarity, enclosure, closure, continuity, and connection. For each, I'll show an example of the principle applied to a table or graph.

Proximity

We tend to think of objects that are physically close together as belonging to part of a group. The proximity principle is demonstrated in Figure 3.1: you naturally see the dots as three distinct groups because of their relative proximity to each other.

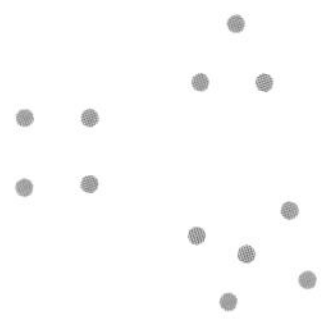

FIGURE 3.1   Gestalt principle of proximity

We can leverage this way that people see in table design. In Figure 3.2, simply by virtue of differentiating the spacing between the dots, your eyes are drawn either down the columns in the first case or across the rows in the second case.

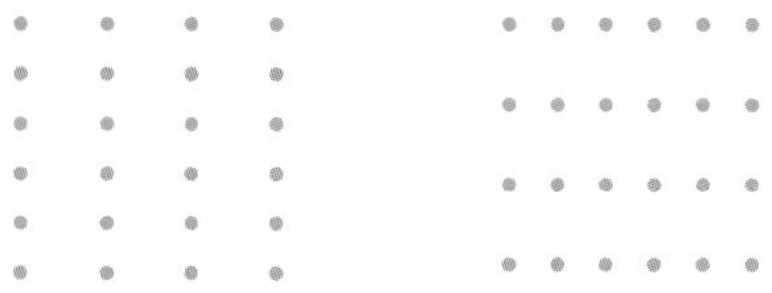

FIGURE 3.2   You see columns and rows, simply due to dot spacing

Similarity

Objects that are of similar color, shape, size, or orientation are perceived as related or belonging to part of a group. In Figure 3.3, you naturally associate the blue circles together on the left or the grey squares together on the right.

FIGURE 3.3   Gestalt principle of similarity

This can be leveraged in tables to help draw our audience's eyes in the direction we want them to focus. In Figure 3.4, the similarity of color is a cue for our eyes to read across the rows (rather than down the columns). This eliminates the need for additional elements such as borders to help direct our attention.

FIGURE 3.4   You see rows due to similarity of color

Enclosure

We think of objects that are physically enclosed together as belonging to part of a group. It doesn't take a very strong enclosure to do this: light background shading is often enough, as demonstrated in Figure 3.5.

FIGURE 3.5   Gestalt principle of enclosure

One way we can leverage the enclosure principle is to draw a visual distinction within our data, as is done in the graph in Figure 3.6.

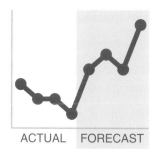

ACTUAL   FORECAST

FIGURE 3.6   The shaded area separates the forecast from actual data

Closure

The closure concept says that people like things to be simple and to fit in the constructs that are already in our heads. Because of this, people tend to perceive a set of individual elements as a single, recognizable shape when they can—when parts of a whole are missing, our eyes fill in the gap. For example, the elements in Figure 3.7 will tend to be perceived as a circle first and only after that as individual elements.

FIGURE 3.7   Gestalt principle of closure

It is common for graphing applications (for example, Excel) to have default settings that include elements like chart borders and background shading. The closure principle tells us that these are unnecessary—we can remove them and our graph still appears as a cohesive entity. Bonus: when we take away those unnecessary elements, our data stands out more, as shown in Figure 3.8.

FIGURE 3.8   The graph still appears complete without the border and background shading

Continuity

The principle of continuity is similar to closure: when looking at objects, our eyes seek the smoothest path and naturally create continuity in what we see even where it may not explicitly exist. By way of example, in Figure 3.9, if I take the objects (1) and pull them apart, most people will expect to see what is shown next (2), whereas it could as easily be what is shown after that (3).

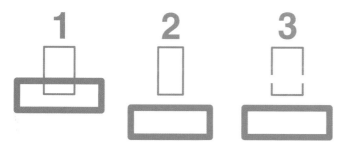

FIGURE 3.9   Gestalt principle of continuity

In the application of this principle, I've removed the vertical y-axis line from the graph in Figure 3.10 altogether. Your eyes actually still see that the bars are lined up at the same point because of the consistent white space (the smoothest path) between the labels on the left and the data on the right. As we saw with the closure principle in application, stripping away unnecessary elements allows our data to stand out more.

FIGURE 3.10   Graph with y-axis line removed

Connection

The final Gestalt principle we'll focus on is connection. We tend to think of objects that are physically connected as part of a group. The connective property typically has a stronger associative value than similar color, size, or shape. Note when looking at Figure 3.11, your eyes probably pair the shapes connected by lines (rather than similar color, size, or shape): that's the connection principle in action. The connective property *isn't* typically stronger than enclosure, but you can impact this relationship through thickness and darkness of lines to create the desired visual hierarchy (we'll talk more about visual hierarchy when we discuss preattentive attributes in Chapter 4).

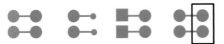

FIGURE 3.11   Gestalt principle of connection

One way that we frequently leverage the connection principle is in line graphs, to help our eyes see order in the data, as shown in Figure 3.12.

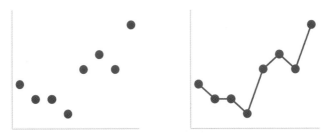

FIGURE 3.12   Lines connect the dots

As you have learned from this brief overview, the Gestalt principles help us understand how people see, which we can use to identify unnecessary elements and ease the processing of our visual communications. We aren't done with them yet. At the end of this chapter, we'll discuss how we can apply some of these principles to a real-world example.

But first, let's shift our focus to a couple of other types of visual clutter.

Lack of visual order

When design is thoughtful, it fades into the background so that your audience doesn't even notice it. When it's not, however, your audience feels the burden. Let's look at an example to understand the impact visual order—and lack thereof—can have on our visual communications.

Take a moment to study Figure 3.13, which summarizes survey feedback about factors considered by nonprofits in vendor selection. Note specifically any observations you may have regarding the arrangement of elements on the page.

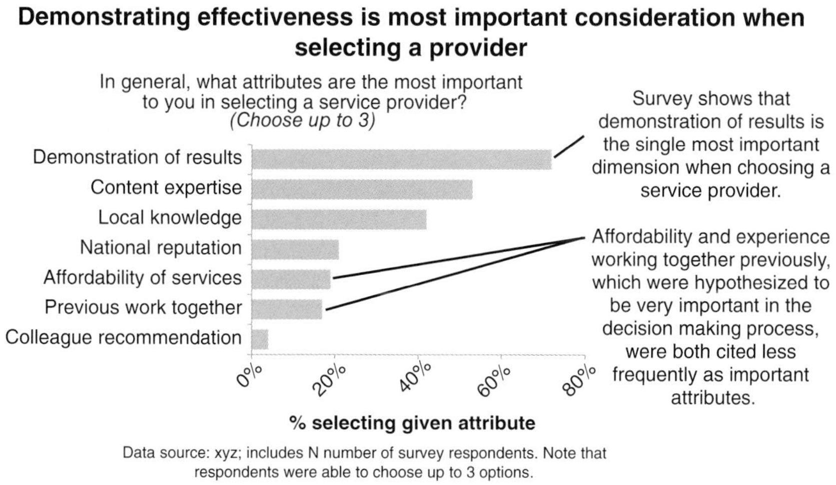

Demonstrating effectiveness is most important consideration when selecting a provider

In general, what attributes are the most important to you in selecting a service provider? *(Choose up to 3)*

Demonstration of results
Content expertise
Local knowledge
National reputation
Affordability of services
Previous work together
Colleague recommendation

0%   20%   40%   60%   80%

% selecting given attribute

Survey shows that demonstration of results is the single most important dimension when choosing a service provider.

Affordability and experience working together previously, which were hypothesized to be very important in the decision making process, were both cited less frequently as important attributes.

Data source: xyz; includes N number of survey respondents. Note that respondents were able to choose up to 3 options.

FIGURE 3.13   Summary of survey feedback

As you look over the information, you might be thinking, "this looks pretty good." I'll concede: it's not horrible. On the positive side, the takeaway is clearly outlined, the graph is well ordered and labeled, and key observations are articulated and tied visually to where we're meant to look in the graph. But when it comes to the overall design of the page and placement of elements, I'd have to disagree with any praise. To me, the aggregate visual feels disorganized and uncomfortable to look at, as if the various components were haphazardly put there without regard for the structure of the overall page.

We can improve this visual markedly by making some relatively minor changes. Take a look at Figure 3.14. The content is exactly the same; only the placement and formatting of elements have been modified.

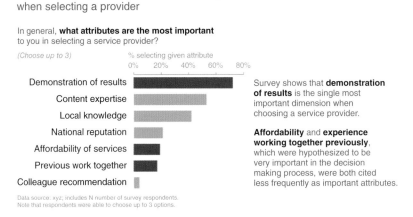

FIGURE 3.14   Revamped summary of survey feedback

Compared to the original visual, the second iteration feels somehow easier. There is order. It is evident that conscious thought was paid to the overarching design and arrangement of components. Specifically, the latter version has been designed with greater attention to alignment and white space. Let's look at each of these in detail.

Alignment

The single change having the biggest impact in the preceding before-and-after example was the shift from center-aligned to left-justified text. In the original version, each block of text on the page is center-aligned. This does not create clean lines either on the left or on the right, which can make even a thoughtful layout appear sloppy. I tend to avoid center-aligned text for this reason. The decision of whether to left- or right-justify your text should be made in context of the other elements on the page. In general, the goal is to create clean lines (both horizontally and vertically) of elements and white space.

Presentation software tips for aligning elements

To help ensure that your elements line up when you are placing them on a page within your presentation software, turn on the rulers or gridlines that are built into most programs. This will allow you to precisely align your elements to create a cleaner look and feel. The table functionality built into most presentation applications can also be used as a makeshift brute-force method: create a table to give yourself guidelines for the placement of discrete elements. When you have everything lined up exactly like you want it, remove the table or make the table's borders invisible so that all that is left is your perfectly arranged page.

Without other visual cues, your audience will typically start at the top left of the page or screen and will move their eyes in a "z" shape (or multiple "z" shapes, depending on the layout) across the page or screen as they take in information. Because of this, when it comes to tables and graphs, I like to upper-left-most justify the text (title, axis titles, legend). This means the audience will hit the details that tell them how to read the table or graph before they get to the data itself.

As part of our discussion on alignment, let's spend a bit of time on **diagonal components.** In the previous example, the original version (Figure 3.13) had diagonal lines connecting the takeaways to the data and diagonally oriented x-axis labels; the former were removed and the latter changed to horizontal orientation in the makeover (Figure 3.14). Generally, diagonal elements such as lines and text should be avoided. They look messy and, in the case of text, are harder to read than their horizontal counterparts. When it comes to the orientation of text, one study (Wigdor & Balakrishnan, 2005) found that the reading of rotated text 45 degrees in either direction was, on average, 52% slower than reading normally oriented text (text rotated 90 degrees in either direction was 205% slower on average). It is best to avoid diagonal elements on the page.

White space

I've never quite understood this phenomenon, but for some reason, people tend to fear white space on a page. I use "white space" to refer to blank space on the page. If your pages are blue, for example, this would be "blue space"—I'm not sure why they would be blue, but the use of color is a conversation we will have later. Perhaps you've heard this feedback before: "there is still some space left on that page, so let's add something there," or worse, "there is still some space left on that page, so let's add more data." No! Never add data just for the sake of adding data—only add data with a thoughtful and specific purpose in mind!

We need to get more comfortable with white space.

White space in visual communication is as important as *pauses* in public speaking. Perhaps you have sat through a presentation that lacked pauses. It feels something like this: *there is a speaker up in front of you and possibly due to nerves or perhaps because they're trying to get through more material than they should in the allotted time they are speaking a mile a minute and you're wondering how they're even able to breathe you'd like to ask a question but the speaker has already moved on to the next topic and still hasn't paused long enough for you to be able to raise your question.* This is an uncomfortable experience for the audience, similar to the discomfort you may have felt reading through the preceding run-on, unpunctuated sentence.

Now imagine the effect if that same presenter were to make a single bold statement: "Death to pie charts!"

And then pause for a full 15 seconds to let that statement resonate.

Go ahead—say it out loud and then count to 15 slowly.

That's a dramatic pause.

And it got your attention, didn't it?

That is the same powerful effect that white space used strategically can have on our visual communications. The lack of it—like the lack of pauses in a spoken presentation—is simply uncomfortable for our audience. Audience discomfort in response to the design of our visual communications is something we should aim to avoid. White space can be used strategically to draw attention to the parts of the page that are *not* white space.

When it comes to preserving white space, here are some minimal guidelines. Margins should remain free of text and visuals. Resist the urge to stretch visuals to take up the available space; instead, appropriately size your visuals to their content. Beyond these guidelines, think about how you can use white space strategically for emphasis, as was illustrated with the dramatic pause earlier. If there is one thing that is really important, think about making that *the only thing on the page*. In some cases, this could be a single sentence or even a single number. We'll talk further about using white space strategically and look at an example when we discuss aesthetics in Chapter 5.

Non-strategic use of contrast

Clear contrast can be a signal to our audience, helping them understand where to focus their attention. We will further explore this idea in greater detail in later chapters. The *lack of clear contrast*, on the other hand, can be a form of visual clutter. When discussing the critical value of contrast, there is an analogy I often borrow from Colin Ware (*Information Visualization: Perception for Design*, 2004), who said it's easy to spot a hawk in a sky full of pigeons, but as the variety of birds increases, that hawk becomes harder and harder to pick out. This highlights the importance of the strategic use of contrast in visual design: the more things we make different, the lesser the degree to which any of them stand out. To explain this another way, if there is something really important we want our audience to know or see (the hawk), we should make that *the one thing* that is very different from the rest.

Let's look at an example to further illustrate this concept.

Imagine you work for a U.S. retailer and want to understand how your customers feel about various dimensions of their shopping experience in your store compared to your competitors. You have conducted a survey to collect this information and are now trying

to understand what it tells you. You have created a weighted performance index to summarize each category of interest (the higher
the index, the better the performance, and vice versa). Figure 3.15
shows the weighted performance index across categories for your
company and five competitors.

Study it for a moment and make note of your thought process as
you take in the information.

FIGURE 3.15   Original graph

If you had to describe Figure 3.15 in a single word, what would that
word be? Words like *busy, confusing*, and perhaps *exhausting* come
to mind. There is a lot going on in this graph. So many things are
competing for our attention that it is hard to know where to look.

Let's review exactly what we're looking at. As I mentioned, the data
graphed is a weighted performance index. You don't need to worry
about the details of how this is calculated, but rather understand
that this is a summary performance metric that we'd like to compare across various categories (shown across the horizontal x-axis:
Selection, Convenience, Service, Relationship, and *Price*) for "Our
Business" (depicted by the blue diamond) compared to a number

of competitors (the other colored shapes). A higher index represents better performance, and a lower index means lower performance.

Taking in this information is a slow process, with a lot of back and forth between the legend at the bottom and the data in the graph to decipher what is being conveyed. Even if we are very patient and really want to get information out of this visual, it is nearly impossible because "Our Business" (the blue diamond) is sometimes obscured by other data points, making it so we can't even see the comparison that is most important to make!

This is a case where lack of contrast (as well as some other design issues) makes the information much harder to interpret than it need be.

Consider Figure 3.16, where we use contrast more strategically.

FIGURE 3.16   Revamped graph, using contrast strategically

In the revised graph, I've made a number of changes. First, I chose a horizontal bar chart to depict the information. In doing so, I rescaled all the numbers to be on a positive scale—in the original scatterplot, there were some negative values that complicated the visualization challenge. This change works here since we're more interested in relative differences than absolute values. In this remake, the categories that were previously along the horizontal x-axis now run down the vertical y-axis. Within each category, the length of the bar shows the summary metric across "Our business" (blue) and the various competitors (grey), with longer bars representing better performance. The decision not to show the actual x-axis scale in this case was a deliberate one, which forces the audience to focus on relative differences rather than get caught up in the minutiae of the specific numbers.

With this design, it is easy to see two things quickly:

1. We can let our eyes scan across the blue bars to get a relative sense of how "Our business" is doing across the various categories: we score high on Price and Convenience and lower on Relationship, possibly because we're struggling when it comes to Service and Selection, as evidenced by low scores in these areas.

2. Within a given category, we can compare the blue bar to the grey bars to see how our business is faring relative to competitors: winning compared to the competition on Price, losing on Service and Selection.

Competitors are distinguished from each other based on the order in which they appear (Competitor A always appears directly after the blue bar, Competitor B after that, and so on), which is outlined in the legend at the left. If it were important to be able to quickly identify each competitor, this design doesn't immediately allow for that. But if that is a second- or third-order comparison in terms of priority and isn't the most critical thing, this approach can work well. In the makeover, I've also organized the categories in order of decreasing weighted performance index for "Our business," which provides a construct for our audience to use as they take in the information, and added a summary metric (relative rank) so it's easy to know quickly how "Our business" ranks in each category in relation to our competition.

Note here how the effective use of contrast (and some other thoughtful design choices) makes it a much faster, easier, and just more comfortable-feeling process to get the information we're after than it was in the original graph.

When redundant details shouldn't be considered clutter

I've seen cases where the title of the visual indicates the values are dollars but the dollar signs aren't included with the actual numbers in the table or graph. For example, a graph titled "Monthly Sales ($USD Millions)" with y-axis labels of 10, 20, 30, 40, 50. I find this confusing. Including the "$" sign with each number eases the interpretation of the figures. Your audience doesn't have to remember they are looking at dollars because they are labeled explicitly. There are some elements that should always be retained with numbers, including dollar signs, percent signs, and commas in large numbers.

Decluttering: step-by-step

Now that we have discussed what clutter is, why it is important to eliminate it from our visual communications, and how to recognize it, let's look at a real-world example and examine how the process of identifying and removing clutter improves our visual and the clarity of the story that we're ultimately trying to tell.

Scenario: Imagine that you manage an information technology (IT) team. Your team receives tickets, or technical issues, from employees. In the past year, you've had a couple of people leave and decided at the time not to replace them. You have heard a rumbling of complaints from the remaining employees about having to "pick up the slack." You've just been asked about your hiring needs for the coming year and are wondering if you should hire a couple more people. First, you want to understand what impact the departure of individuals over the past year has had on your team's overall productivity. You plot the monthly trend of incoming tickets and those processed over the past calendar year. You see that there is some evidence your team's productivity is suffering from being short-staffed and now want to turn the quick-and-dirty visual you created into the basis for your hiring request.

Figure 3.17 shows your original graph.

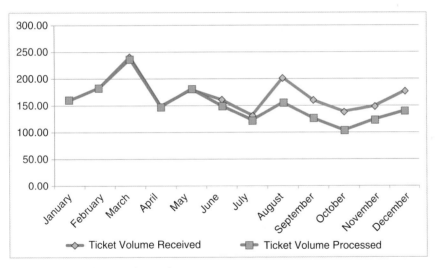

FIGURE 3.17 Original graph

Take another look at this visual with an eye toward clutter. Consider the lessons we've covered on Gestalt principles, alignment, white space, and contrast. What things can we get rid of or change? How many issues can you identify?

I identified six major changes to reduce clutter. Let's discuss each.

1. Remove chart border

Chart borders are usually unnecessary, as we covered in our discussion of the Gestalt principle of closure. Instead, think about using white space to differentiate the visual from other elements on the page as needed.

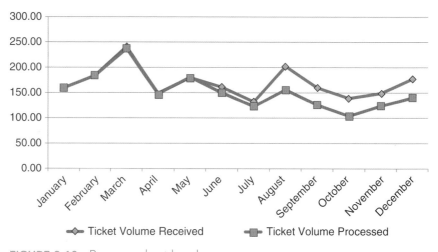

FIGURE 3.18   Remove chart border

2. Remove gridlines

If you think it will be helpful for your audience to trace their finger from the data to the axis, or you feel that your data will be more effectively processed, you can leave the gridlines. But make them thin and use a light color like grey. Do not let them compete visually with your data. When you can, get rid of them altogether: this allows for greater contrast, and your data will stand out more.

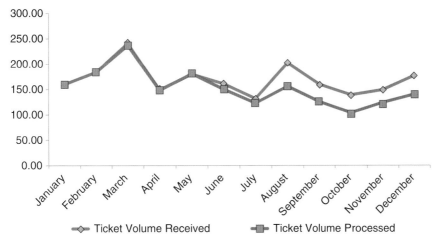

FIGURE 3.19   Remove gridlines

3. Remove data markers

Remember, every single element adds cognitive load on the part of your audience. Here, we're adding cognitive load to process data that is already depicted visually with the lines. This isn't to say that you should never use data markers, but rather use them on purpose and with a purpose, rather than because their inclusion is your graphing application's default.

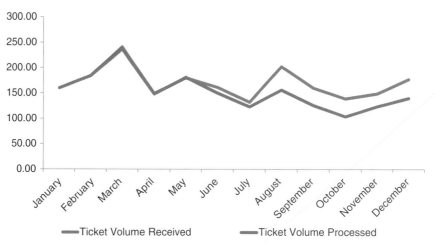

FIGURE 3.20   Remove data markers

4. Clean up axis labels

One of my biggest pet peeves is trailing zeros on y-axis labels: they carry no informative value, and yet make the numbers look more complicated than they are! Get rid of them, reducing their unnecessary burden on the audience's cognitive load. We can also abbreviate the months of the year so that they will fit horizontally on the x-axis, eliminating the diagonal text.

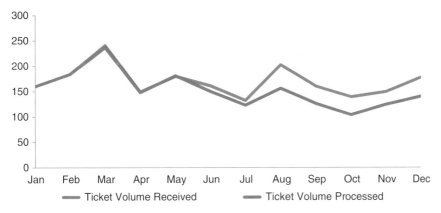

FIGURE 3.21   Clean up axis labels

5. Label data directly

Now that we have eliminated much of the extraneous cognitive load, the work of going back and forth between the legend and the data is even more evident. Remember, we want to try to identify anything that will feel like effort to our audience and take that work upon ourselves as the designers of the information. In this case, we can leverage the Gestalt principle of proximity and put the data labels right next to the data they describe.

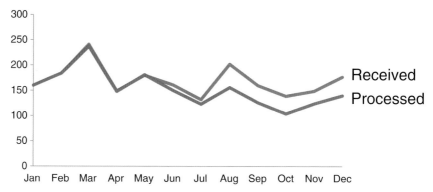

FIGURE 3.22   Label data directly

6. Leverage consistent color

While we leveraged the Gestalt principle of proximity in the prior step, let's also think about leveraging the Gestalt principle of similarity and make the data labels the same *color* as the data they describe. This is another visual cue to our audience that says, "these two pieces of information are related."

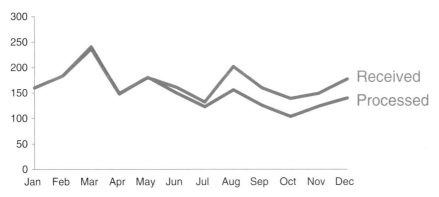

FIGURE 3.23   Leverage consistent color

This visual is not yet complete. But identifying and eliminating the clutter has brought us a long way in terms of reducing cognitive load and improving accessibility. Take a look at the before-and-after shown in Figure 3.24.

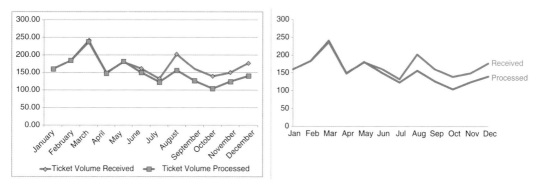

FIGURE 3.24   Before-and-after

In closing

Any time you put information in front of your audience, you are creating cognitive load and asking them to use their brain power to process that information. Visual clutter creates excessive cognitive load that can hinder the transmission of our message. The Gestalt Principles of Visual Perception can help you understand how your audience sees and allow you to identify and remove unnecessary visual elements. Leverage alignment of elements and maintain white space to help make the interpretation of your visuals a more comfortable experience for your audience. Use contrast strategically. Clutter is your enemy: ban it from your visuals!

You now know how to **identify and eliminate clutter**.

chapter four

focus your audience's attention

In the previous chapter, we learned about clutter and the importance of identifying and removing it from our visuals. While we work to eliminate distractions, we also want to look at what remains and consider how we want our audience to interact with our visual communications.

In this chapter, we further examine how people see and how you can use that to your advantage when crafting visuals. We will talk briefly about sight and memory in order to highlight the importance of some specific, powerful tools: **preattentive attributes**. We will explore how preattentive attributes like size, color, and position on page can be used strategically in two ways. First, preattentive attributes can be leveraged to help direct your audience's attention to where you want them to focus it. Second, they can be used to create a visual hierarchy of elements to lead your audience through the information you want to communicate in the way you want them to process it.

By understanding how our audience sees and processes information, we put ourselves in a better position to be able to communicate effectively.

You see with your brain

Let's look at a simplified picture of how people see, depicted in Figure 4.1. The process goes something like this: light reflects off of a stimulus. This gets captured by our eyes. We don't fully see with our eyes; there is some processing that happens there, but mostly it is what happens in our brain that we think of as visual perception.

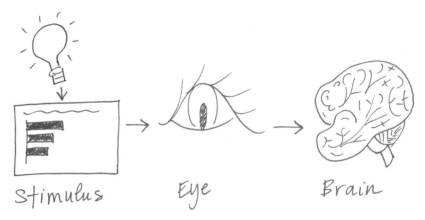

FIGURE 4.1 A simplified picture of how you see

A brief lesson on memory

Within the brain, there are three types of memory that are important to understand as we design visual communications: iconic memory, short-term memory, and long-term memory. Each plays an important and distinct role. What follows are basic explanations of highly complex processes, covered simply to set the stage for what you need to know when designing visual communications.

Iconic memory

Iconic memory is super fast. It happens without you consciously realizing it and is piqued when we look at the world around us. *Why?* Long ago in the evolutionary chain, predators helped our brains develop in ways that allowed for great efficiency of sight and speed of response. In particular, the ability to quickly pick up differences in our environment—for example, the motion of a predator in the distance—became ingrained in our visual process. These were survival mechanisms then; they can be leveraged for effective visual communication today.

Information stays in your iconic memory for a fraction of a second before it gets forwarded on to your short-term memory. The important thing about iconic memory is that it is tuned to a set of preattentive attributes. Preattentive attributes are critical tools in your visual design tool belt, so we'll come back to those in a moment. In the meantime, let's continue our discussion on memory.

Short-term memory

Short-term memory has limitations. Specifically, people can keep about four chunks of visual information in their short-term memory at a given time. This means that if we create a graph with ten different data series that are ten different colors with ten different shapes of data markers and a legend off to the side, we're making our audience work very hard going back and forth between the legend and the data to decipher what they are looking at. As we've discussed previously, to the extent possible, we want to limit this sort of cognitive burden on our audience. We don't want to make our audience work to get at the information, because in doing so, we run the risk of losing their attention. With that, we lose our ability to communicate.

In this specific situation, one solution is to label the various data series directly (reducing that work of going back and forth between the legend and the data by leveraging the Gestalt principle of proximity that we covered in Chapter 3). More generally, we want to form larger, coherent chunks of information so that we can fit them into the finite space in our audience's working memory.

Long-term memory

When something leaves short-term memory, it either goes into oblivion and is likely lost forever, or is passed into long-term memory. Long-term memory is built up over a lifetime and is vitally important for pattern recognition and general cognitive processing. Long-term memory is the aggregate of visual and verbal memory, which act differently. Verbal memory is accessed by a neural net, where the path becomes important for being able to recognize or recall. Visual memory, on the other hand, functions with specialized structures.

There are aspects of long-term memory that we want to make use of when it comes to having our message stick with our audience. Of particular importance to our conversation is that images can help us more quickly recall things stored in our long-term verbal memory. For example, if you see a picture of the Eiffel Tower, a flood of concepts you know about, feelings you have toward, or experiences you've had in Paris may be triggered. By combining the visual and verbal, we set ourselves up for success when it comes to triggering the formation of long-term memories in our audience. We'll discuss some specific tactics for this in Chapter 7 in the context of storytelling.

Preattentive attributes signal where to look

In the previous section, I introduced iconic memory and mentioned that it is tuned to preattentive attributes. The best way to prove the power of preattentive attributes is to demonstrate it. Figure 4.2 shows a block of numbers. Taking note of how you process the information and how long it takes, quickly count the number of 3s that appear in the sequence.

75639506847 3
65866303757 6
86037265860 2
84658910783 0

FIGURE 4.2   Count the 3s example

The correct answer is six. In Figure 4.2, there were no visual cues to help you reach this conclusion. This makes for a challenging exercise, during which you have to hunt through four lines of text, looking for the number 3 (a kind of complicated shape).

Check out what happens when we make a single change to the block of numbers. Turn the page and repeat the exercise of counting the 3s using Figure 4.3.

756**3**95068473**3**
65866**3**0**3**7576
860**3**72658602
84658910783**0**

FIGURE 4.3   Count the 3s example with preattentive attributes

Note how much easier and faster the same exercise is using Figure 4.3. You don't have time to blink, don't really have time to think, and suddenly there are six 3s in front of you. This is so apparent so quickly because in this second iteration, your iconic memory is being leveraged. The preattentive attribute of intensity of color, in this case, makes the 3s the one thing that stands out as distinct from the rest. Our brain is quick to pick up on this without our having to dedicate any conscious thought to it.

This is remarkable. And profoundly powerful. It means that, if we use preattentive attributes strategically, they can help us **enable our audience to see what we want them to see before they even know they're seeing it!**

Note the multiple preattentive attributes I've used in the preceding text to underscore its importance!

Figure 4.4 shows the various preattentive attributes.

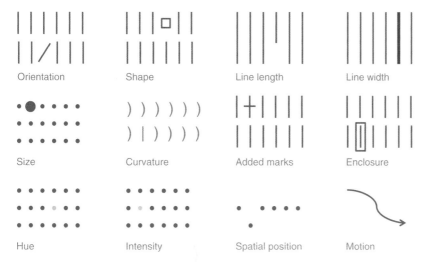

FIGURE 4.4   Preattentive attributes

Source: Adapted from Stephen Few's *Show Me the Numbers*, 2004.

Note as you scan across the attributes in Figure 4.4, your eye is drawn to the one element within each group that is different from the rest: you don't have to look for it. That's because our brains are hardwired to quickly pick up differences we see in our environment.

One thing to be aware of is that people tend to associate quantitative values with some (but not all) of the preattentive attributes. For example, most people will consider a long line to represent a greater value than a short line. That is one of the reasons bar charts are straightforward for us to read. But we don't think of color in the same way. If I ask you which is greater—red or blue?—this isn't a meaningful question. This is important because it tells us which of the attributes can be used to encode quantitative information (line length, spatial position, or to a more limited extent, line width, size, and intensity can be used to reflect relative value), and which should be used as categorical differentiators.

When used sparingly, preattentive attributes can be extremely useful for doing two things: (1) drawing your audience's attention quickly to where you want them to look, and (2) creating a visual hierarchy of information. Let's look at examples of each of these, first with text and then in the context of data visualization.

Preattentive attributes in text

Without any visual cues, when we're confronted with a block of text, our only option is to read it. But preattentive attributes employed sparingly can quickly change this. Figure 4.5 shows how you can utilize some of the preattentive attributes introduced previously with text. The first block of text doesn't employ any preattentive attributes. This renders it similar to the count the 3s example: you have to read it, put on the lens of what's important or interesting, then possibly read it again to put the interesting parts back into the context of the rest.

Observe how leveraging preattentive attributes changes the way you process the information. The subsequent blocks of text employ a single preattentive attribute each. Note how, within each, the preattentive attribute grabs your attention, and how some attributes draw your eyes with greater or weaker force than others (for example, color and size are attention grabbing, whereas italics achieve a milder emphasis).

No preattentive attributes

What are we doing well? Great Products. These products are clearly the best in their class. Replacement parts are shipped when needed. You sent me gaskets without me having to ask. Problems are resolved promptly. Bev in the billing office was quick to resolve a billing issue I had. General customer service exceeds expectations. The account manager even called to check in after normal business hours.
You have a great company – keep up the good work!

Bold

What are we doing well? Great Products. These products are clearly the best in their class. Replacement parts are shipped when needed. You sent me gaskets without me having to ask. Problems are resolved promptly. Bev in the billing office was quick to resolve a billing issue I had. General customer service exceeds expectations. The account manager even called to check in after normal business hours.
You have a great company – keep up the good work!

Color

What are we doing well? Great Products. **These products are clearly the best in their class.** Replacement parts are shipped when needed. You sent me gaskets without me having to ask. Problems are resolved promptly. Bev in the billing office was quick to resolve a billing issue I had. General customer service exceeds expectations. The account manager even called to check in after normal business hours.
You have a great company – keep up the good work!

Italics

What are we doing well? Great Products. These products are clearly the best in their class. *Replacement parts are shipped when needed*. You sent me gaskets without me having to ask. Problems are resolved promptly. Bev in the billing office was quick to resolve a billing issue I had. General customer service exceeds expectations. The account manager even called to check in after normal business hours.
You have a great company – keep up the good work!

Size

What are we doing well? Great Products. These products are the best in their class. Replacement parts are shipped when needed. You sent gaskets

without me having to ask. Problems are resolved promptly. Bev in the

billing office was quick to resolve a billing issue I had. General customer service exceeds expectations. The account manager even called to check in after normal business hours. You have a great company – keep up the good work!

Separate spatially

What are we doing well? Great Products. These products are clearly the best in their class. Replacement parts are shipped when needed. You sent me gaskets without me having to ask.

Problems are resolved promptly.

Bev in the billing office was quick to resolve a billing issue I had. General customer service exceeds expectations. The account manager even called to check in after normal business hours. You have a great company – keep up the good work!

Outline (enclosure)

What are we doing well? Great Products. These products are clearly the best in their class. Replacement parts are shipped when needed. You sent me gaskets without me having to ask. Problems are resolved promptly. Bev in the billing office was quick to resolve a billing issue I had. General customer service exceeds expectations. The account manager even called to check in after normal business hours.
You have a great company – keep up the good work!

Underline (added marks)

What are we doing well? Great Products. These products are clearly the best in their class. Replacement parts are shipped when needed. You sent me gaskets without me having to ask. Problems are resolved promptly. Bev in the billing office was quick to resolve a billing issue I had. General customer service exceeds expectations. The account manager even called to check in after normal business hours.
<u>You have a great company – keep up the good work!</u>

FIGURE 4.5   Preattentive attributes in text

Beyond drawing our audience's attention to where we want them to focus it, we can employ preattentive attributes to create **visual hierarchy** in our communications. As we saw in Figure 4.5, the various attributes draw our attention with differing strength. In addition, there are variances within a given preattentive attribute that will draw attention with more or less strength. For example, with the preattentive attribute of color, a bright blue will typically draw attention more than a muted blue. Both will draw more attention than a light grey. We can leverage this variance and use multiple preattentive attributes together to make our visuals scannable, by emphasizing some components and de-emphasizing others.

Figure 4.6 illustrates how this can be done with the block of text from the previous example.

What are we doing well?
Themes & example comments

- **Great products**: "These products are clearly the best in class."
- **Replacement parts are shipped when needed**:
  "You sent me gaskets without me having to ask, and I really needed them, too!"
- **Problems are resolved promptly**: "Bev in the billing office was quick to resolve a billing issue I had."
- **General customer service exceeds expectations**:
  "The account manager even called after normal business hours. You have a great company - keep up the good work!"

FIGURE 4.6   Preattentive attributes can help create a visual hierarchy of information

Preattentive attributes have been used in Figure 4.6 to create a visual hierarchy of information. This makes the information we present more easily scannable. Studies have shown that we have about 3–8 seconds with our audience, during which time they decide whether to continue to look at what we've put in front of them or direct their attention to something else. If we've used our preattentive attributes wisely, even if we only get that initial 3–8 seconds, we've given our audience the gist of what we want to say.

Leveraging preattentive attributes to create a clear visual hierarchy of information establishes implicit instructions for your audience, indicating to them how to process the information. We can signal what is most important that they should pay attention to first, what is second most important that they should pay attention to next, and so on. We can push necessary but non-message-impacting components to the background so they don't compete for attention. This makes it both easier and faster for our audience to take in the information that we provide.

The preceding example demonstrated the use of preattentive attributes in text. Preattentive attributes are also very useful for communicating effectively with data.

Preattentive attributes in graphs

Graphs, without other visual cues, can become very much like the count the 3s exercise or the block of text we've considered previously. Take the following example. Imagine you work for a car manufacturer. You are interested in understanding and sharing insight about the top design concerns (measured as the number of concerns per 1,000 concerns) from customers for a particular vehicle make and model. Your initial visual might look something like Figure 4.7.

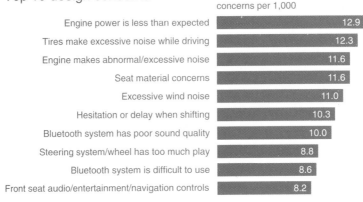

Top 10 design concerns

FIGURE 4.7   Original graph, no preattentive attributes

Note how, without other visual cues, you are left to process *all of the information*. With no clues about what's important or should be paid attention to, it's the count the 3s exercise all over again.

Recall the distinction that was drawn early on in Chapter 1 between exploratory and explanatory analysis. The visual in Figure 4.7 could be one you create during the exploratory phase: when you're looking at the data to understand what might be interesting or noteworthy to communicate to someone else. Figure 4.7 shows us that there are ten design concerns that have more than eight concerns per 1,000.

When it comes to explanatory analysis and leveraging this visual to share *information* with your audience (rather than just showing data), thoughtful use of color and text is one way we can focus the story, as illustrated in Figure 4.8.

7 of the top 10 design concerns have 10 or more concerns per 1,000.
Discussion: is this an acceptable default rate?

Top 10 design concerns

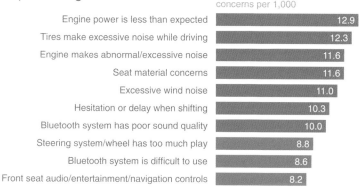

FIGURE 4.8   Leverage color to draw attention

We can go one step further, using the same visual but with modified focus and text to lead our audience from the macro to the micro parts of the story, as demonstrated in Figure 4.9.

Of the top design concerns, three are noise-related.

Top 10 design concerns

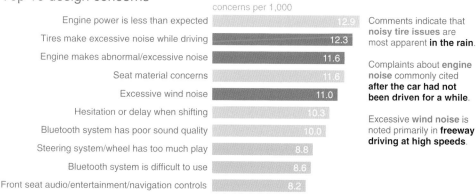

FIGURE 4.9   Create a visual hierarchy of information

Especially in live presentation settings, repeated iterations of the same visual, with different pieces emphasized to tell different stories or different aspects of the same story (as demonstrated in Figures 4.7,

4.8, and 4.9), can be an effective strategy. This allows you to familiarize your audience with your data and visual first and then continue to leverage it in the manner illustrated. Note in this example how your eyes are drawn to the elements of the visual you're meant to focus on due to strategic use of preattentive attributes.

Highlighting one aspect can make other things harder to see

One word of warning in using preattentive attributes: when you highlight one point in your story, it can actually make other points harder to see. When you're doing exploratory analysis, you should mostly avoid the use of preattentive attributes for this reason. When it comes to *explanatory* analysis, however, you should have a specific story you are communicating to your audience. Leverage preattentive attributes to help make that story visually clear.

The previous example used mainly color to draw the viewer's attention. Let's look at another scenario using a different preattentive attribute. Recall the example introduced in Chapter 3: you manage an IT team and want to show how the volume of incoming tickets exceeds your team's resources. After decluttering the graph, we were left with Figure 4.10.

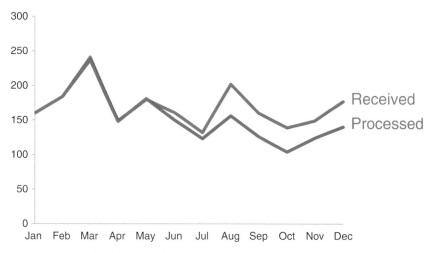

FIGURE 4.10 Let's revisit the ticket example

In the process of determining where I want to focus my audience's attention, one strategy I'll often employ is to start by pushing *everything* to the background. This forces me to make explicit decisions regarding what to bring to the forefront or highlight. Let's start by doing this; see Figure 4.11.

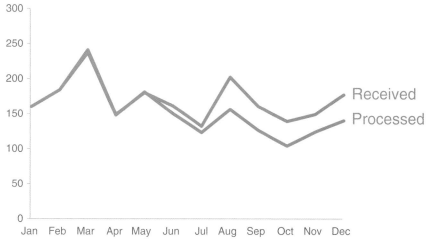

FIGURE 4.11 First, push everything to the background

Next, I want to make the data stand out. Figure 4.12 shows both data series (Received and Processed) bolder and bigger than axis lines and labels. It was an intentional decision to make the Processed line darker than the Received line to draw emphasis to the fact that the number of tickets being processed has fallen below the number being received.

FIGURE 4.12   Make the data stand out

In this case, we want to draw our audience's attention to the right side of the graph, where the gap has started to form. Without other visual cues, our audience will typically start at the top left of our visual and do zigzagging "z's" with their eyes across the page. The viewer will eventually get to that gap on the right-hand side, but let's consider how we can use our preattentive attributes to make that happen more quickly.

The added marks of data points and numeric labels are one preattentive attribute we can leverage. Bear with me, though, as we take a step in the wrong direction before we go in the right one. See Figure 4.13.

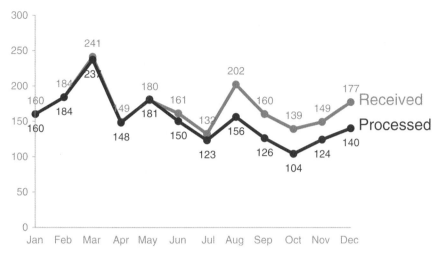

FIGURE 4.13   Too many data labels feels cluttered

When we add data markers and numeric labels to every data point, we quickly create a cluttered mess. But check out what happens in Figure 4.14 when we're strategic about *which* data markers and labels we preserve and which we eliminate.

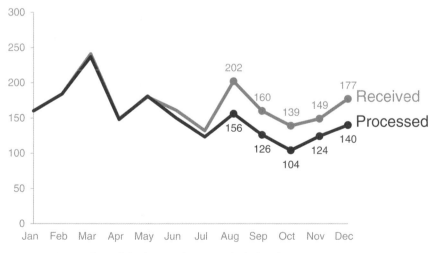

FIGURE 4.14   Data labels used sparingly help draw attention

In Figure 4.14, the added marks act as a "look here" signal, drawing our audience's attention more quickly to the right side of the graph.

They provide for our audience the added benefit of allowing them to do some quick math in the event that they want to understand how big the backlog is becoming (if we think that is something they'd definitely want to do, we should consider doing it for them).

These are just a couple of examples of using preattentive attributes to focus the audience's attention. We will look at a number of additional examples that leverage this same broad strategy in different ways throughout the rest of this book.

There are a few preattentive attributes that are so important from a strategic standpoint when it comes to focusing your audience's attention that they warrant their own specific discussions: size, color, and position on page. We'll address each of these in the following sections.

Size

Size matters. Relative size denotes relative importance. Keep this in mind when designing your visual communications. If you're showing multiple things that are of roughly equal importance, size them similarly. Alternatively, if there is one really important thing, leverage size to indicate that: make it BIG!

The following is a real situation where size nearly caused unintended repercussions.

Early in my career at Google, we were designing a dashboard to help with a decision-making process (I'm being intentionally vague to preserve confidentiality). In the design phase, there were three main pieces of information we knew we wanted to include, only one of which was readily available (the other data had to be chased after). In the initial versions of the dashboard, the information we had on hand took up probably 60% of the dashboard's real estate, with placeholders for the other information we were collecting. After getting our hands on the other data, we plugged it into the existing placeholders. Rather late in the game, we realized that the size of that initial data we had included was drawing undue attention compared to the rest of the information on the page. Luckily, we caught this before it was too late. We modified the layout to make the three equally important things the same size. It's interesting to think that completely different conversations may have been had and decisions reached as a result of this shift in design.

This was an important lesson for me (and one that we'll highlight in the next section on color as well): don't let your design choices be happenstance; rather, they should be the result of explicit decisions.

Color

When used sparingly, color is one of the most powerful tools you have for drawing your audience's attention. Resist the urge to use color for the sake of being colorful; instead, leverage color selectively as a strategic tool to highlight the important parts of your visual. The use of color should always be an intentional decision. Never let your tool make this important decision for you!

I typically design my visuals in shades of grey and pick a single bold color to draw attention where I want it. My base color is grey, not black, to allow for greater contrast since color stands out more against grey than black. For my attention-grabbing color, I often use blue for a number of reasons: (1) I like it, (2) you avoid issues of colorblindness that we'll discuss momentarily, and (3) it prints well in black-and-white. That said, blue is certainly not your only option (and you'll see many examples where I deviate from my typical blue for various reasons).

When it comes to the use of color, there are several specific lessons to know: use it sparingly, use it consistently, design with the colorblind in mind, be thoughtful of the tone color conveys, and consider whether to leverage brand colors. Let's discuss each of these in detail.

Use color sparingly

It's easy to spot a hawk in a sky full of pigeons, but as the variety of birds increases, that hawk becomes harder and harder to locate. Remember the adage from Colin Ware that we discussed in the last chapter on clutter? The same principle applies here. For color to be effective, it must be used sparingly. Too much variety prevents anything from standing out. There needs to be sufficient contrast to make something draw your audience's attention.

When we use too many colors together, beyond entering rainbowland, we lose their preattentive value. By way of example, I once encountered a table that showed market rank for a handful of pharmaceutical drugs across a number of different countries, similar to the left-hand side of Figure 4.15. Each rank (1, 2, 3, and so on) was assigned its own color along a rainbow spectrum: 1 = red, 2 = orange, 3 = yellow, 4 = light green, 5 = green, 6 = teal, 7 = blue, 8 = dark blue, 9 = light purple, 10+ = purple. The cells within the table were filled with the color that corresponded to the numerical ranking. Rainbow Brite might have loved this table (for those unfamiliar, a quick Google image search of Rainbow Brite will bring some understanding to this statement), but I was not a fan. The power of the preattentive attributes was lost: everything was different, which meant that nothing stood out. We were back to the count the 3s example—only worse, because the variance in colors was actually more distracting than helpful. A better alternative would be to use varying color saturation of a single color (a heatmap).

Country Level Sales Rank Top 5 Drugs

Rainbow distribution in color indicates sales rank in given country from #1 (red) to #10 or higher (dark purple)

Country	A	B	C	D	E
AUS	1	2	3	6	7
BRA	1	3	4	5	6
CAN	2	3	6	12	8
CHI	1	2	8	4	7
FRA	3	2	4	8	10
GER	3	1	6	5	4
IND	4	1	8	10	5
ITA	2	4	10	9	8
MEX	1	5	4	6	3
RUS	4	3	7	9	12
SPA	2	3	4	5	11
TUR	7	2	3	4	8
UK	1	2	3	6	7
US	1	2	4	3	5

Top 5 drugs: country-level sales rank

RANK	1	2	3	4	5+

COUNTRY | DRUG	A	B	C	D	E
Australia	1	2	3	6	7
Brazil	1	3	4	5	6
Canada	2	3	6	12	8
China	1	2	8	4	7
France	3	2	4	8	10
Germany	3	1	6	5	4
India	4	1	8	10	5
Italy	2	4	10	9	8
Mexico	1	5	4	6	3
Russia	4	3	7	9	12
Spain	2	3	4	5	11
Turkey	7	2	3	4	8
United Kingdom	1	2	3	6	7
United States	1	2	4	3	5

FIGURE 4.15   Use color sparingly

Let's consider Figure 4.15. Where are your eyes drawn in the version on the left? Mine dart around quite a bit, trying to figure out what I should pay attention to. They hesitate on the dark purple, then red, then to the dark blue, probably because these have a higher saturation of color than the others. However, when we consider what these colors represent, it's not necessarily where we want our audience to look.

In the version on the right-hand side, varying saturation of a single color is used. Note that our perception is more limited when it comes to relative saturation, but one benefit we get is that it does carry with it some quantitative assumptions (that more heavily saturated represents greater value than less or vice versa—something you don't get with the rainbow colors used originally as categorical differentiators). This works well for our purpose here, where the low numbers (market leaders) are denoted with the highest color saturation. We are drawn to the dark blue first—the market leaders. This is a more thoughtful use of color.

Where are your eyes drawn?

There is an easy test for determining whether preattentive attributes are being used effectively. Create your visual, then close your eyes or look away for a moment and then look back at it, taking note of where your eyes are drawn first. Do they immediately land where you want your audience to focus? Better yet, seek the help of a friend or colleague—ask them to talk you through how they process the visual: where their eyes go first, where they go next, and so on. This is a great way to see things through your audience's eyes and confirm whether the visual you've created is drawing attention and creating a visual hierarchy of information in the way that you desire.

Use color consistently

One question regularly raised in my workshops is around novelty. *Does it make sense to change up the colors or graph types so the audience doesn't get bored?* My answer is a resounding *No!* The story you are telling should be what keeps your audience's attention (we'll talk about story more in Chapter 7), not the design elements of your graphs. When it comes to the type of graph, you should always use whatever will be easiest for your audience to read. When showing similar information that can be graphed the same way, there can be benefit to keeping the same layout as you essentially train your audience how to read the information, making the interpretation of later graphs all the easier and reducing mental fatigue.

A change in colors signals just that—a change. So leverage this when you want your audience to feel change for some reason, but never simply for the sake of novelty. If you are designing your communication in shades of grey and using a single color to draw attention, leverage that same schematic throughout the communication. Your audience quickly learns that blue, for example, signals where they are meant to look first, and can use this understanding as they process subsequent slides or visuals. However, if you want to signal a clear change in topic or tone, a shift in color is one way to visually reinforce this.

There are some cases where use of color must be consistent. Your audience will typically take time to familiarize themselves with what colors mean once and then will assume the same details apply throughout the rest of the communication. For example, if you are displaying data across four regions in a graph, each having their own color in one place within your presentation or report, be sure to preserve this same schematic throughout the visuals in the rest of your presentation or report (and avoid use of the same colors for other purposes if possible). Don't confuse your audience by changing your use of color.

Design with colorblind in mind

Roughly 8% of men (including my husband and a former boss) and half a percent of women are colorblind. This most frequently manifests itself as difficulty in distinguishing between shades of red and shades of green. In general, you should avoid using shades of red and shades of green together. Sometimes, though, there is useful connotation that comes with using red and green: red to denote the double-digit loss you want to draw attention to or green to highlight significant growth. You can still leverage this, but make sure to have some additional visual cue to set the important numbers apart so you aren't inadvertently disenfranchising part of your audience. Consider also using bold, varying saturation or brightness, or adding a simple plus or minus sign in front of the numbers to ensure they stand out.

When I'm designing a visual and selecting colors to highlight both positive and negative aspects, I frequently use blue to signal positive and orange for negative. I feel that positive and negative associations with these colors are still recognizable and you avoid the colorblind challenge described above. When you face this situation, consider whether you need to highlight both ends of the scale (positive and negative) with color, or if drawing attention to one or the other (or sequentially, one and then the other) might work to tell your story.

See your graphs and slides through colorblind eyes

There are a number of sites and applications with colorblindness simulators that allow you to see what your visual looks like through colorblind eyes. For example, *Vischeck* (vischeck.com) allows you to upload images or download the tool to use on your own computer. *Color Oracle* (colororacle.org) offers a free download for Windows, Linux, or Mac that applies a full-screen color filter independent of the software in use. *CheckMyColours* (checkmycolours.com) is a tool for checking foreground and background colors and determining if they provide sufficient contrast when viewed by someone having color-sight deficiency.

Be thoughtful of tone that color conveys

Color evokes emotion. Consider the tone you want to set with your data visualization or broader communication and choose a color (or colors) that help reinforce the emotion you want to arouse from your audience. Is the topic serious or lighthearted? Are you making a striking bold statement and want your colors to echo it, or is a more circumspect approach with a muted color-scheme appropriate?

Let's discuss a couple specific examples of color and tone. I was once told by a client that the visuals I had made over looked "too nice" (as in friendly). I had created these particular visuals in my typical color palette: shades of grey with a medium blue used sparingly to draw attention. They were reporting the results of statistical analysis, and were used to and wanted a more clinical look and feel. Taking this into account, I reworked the visuals to leverage bold black to draw attention. I also swapped some of the title text for all capital letters and changed the font throughout (we'll discuss font in more detail in Chapter 5 in the context of design).

The resulting visuals, though at the core were exactly the same, had a completely different look and feel because of these simple changes. As with many of the other decisions we make when communicating with data, the audience (in this case, my client) should be kept top of mind and their needs and desires considered when making design choices like these.

Cultural color connotations

When picking colors for communications to international audiences, it may be important to consider the connotations colors have in other cultures. David McCandless created a visualization showing colors and what they mean in different cultures, which can be found in his book *The Visual Miscellaneum: A Colorful Guide to the World's Most Consequential Trivia* (2012) or on his website at informationisbeautiful.net/visualizations/colours-in-cultures.

As another example on color and tone, I recall flipping through an airline magazine on a business trip and finding a fluffy article on online dating accompanied by graphs charting related data. The graphs were almost entirely hot pink and teal. Would you choose this color scheme for your quarterly business report? Certainly not. But given the nature and lively tone of the article these visuals accompanied, the peppy colors worked (and caught my attention!).

Brand colors: to leverage or not to leverage?

Some companies go through major undertakings to create their branding and associated color palette. There may be brand colors that you are required to work with or that make sense to leverage. The key to success when that is the case is to identify one or maybe two brand-appropriate colors to use as your "audience-look-here" cues and keep the rest of your color palette relatively muted with shades of grey or black.

In some cases, it may make sense to deviate from brand colors entirely. For example, I was once working with a client whose brand color was a light shade of green. I originally wanted to leverage this green as the standout color, but it simply wasn't attention grabbing enough. There wasn't sufficient contrast, so the visuals I created had a washed-out feel. When this is the case, you can use bold black to draw attention when everything else is in shades of grey, or choose an entirely different color—just make sure it doesn't clash with the brand colors if they need to be shown together (for example, if the brand logo will be on each page of the slide deck you are building). In this particular case, the client favored the version where I used an entirely different color. A sample of each of the approaches is shown in Figure 4.16.

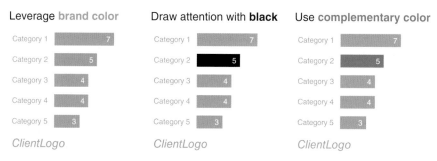

FIGURE 4.16   Color options with brand color

In short: be thoughtful when it comes to your use of color!

Position on page

Without other visual cues, most members of your audience will start at the top left of your visual or slide and scan with their eyes in zig-zag motions across the screen or page. They see the top of the page first, which makes this precious real estate. Think about putting the most important thing here (see Figure 4.17).

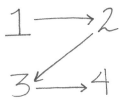

FIGURE 4.17   The zigzag "z" of taking in information on a screen or page

If something is important, try not to make your audience wade through other stuff to get to it. Eliminate this work by putting the important thing at the top. On a slide, these may be words (the main takeaway or call to action). In a data visualization, think about which data you want your audience to see first and whether rearranging the visual accordingly makes sense (it won't always, but this is one tool you have at your disposal for signaling importance to your audience).

Aim to work within the way your audience takes in information, not against it. Here is an example of asking the audience to work against the way that comes naturally to them: I was once shown a process flow diagram that started at the bottom right and you were meant to read it upwards and to the left. This felt really uncomfortable (feelings of discomfort are something we should aim to avoid in our audience!). All I wanted to do was read it from the top left to the bottom right, irrespective of the other visual cues that were present to try to encourage me to do the opposite. Another example I sometimes see in data visualization is something plotted on a scale ranging from negative to positive where the positive values are on the left (which is more typically associated with negative) and the negative values are on the right (which is more naturally associated with positive). Again, in this example, the information is organized in a way that is counter to the way the audience wants to take in the information, rendering the visual challenging to decipher. We'll look at a specific example related to this in case study 3 in Chapter 9.

Be mindful of how you position elements on a page and aim to do so in a way that will feel natural for your audience to consume.

In closing

Preattentive attributes are powerful tools when used sparingly and strategically in visual communication. Without other cues, our audience is left to process *all* of the information we put in front of them. Ease this by leveraging preattentive attributes like size, color, and position on page to signal what's important. Use these strategic attributes to draw attention to where you want your audience to look and create visual hierarchy that helps guide your audience through the visual in the way you want. Evaluate the effectiveness of preattentive attributes in your visual by applying the "where are your eyes drawn?" test.

With that, consider your fourth lesson learned. You now know how to **focus your audience's attention where you want them to pay it**.

chapter five

think like a designer

Form follows function. This adage of product design has clear application to communicating with data. When it comes to the form and function of our data visualizations, we first want to think about what it is we want our audience to be able to *do* with the data (function) and then create a visualization (form) that will allow for this with ease. In this chapter, we will discuss how traditional design concepts can be applied to communicating with data. We will explore **affordances**, **accessibility**, and **aesthetics**, drawing on a number of concepts introduced previously, but looking at them through a slightly different lens. We will also discuss strategies for gaining audience **acceptance** of your visual designs.

Designers know the fundamentals of good design but also how to trust their eye. You may think to yourself, *But I'm not a designer!* Stop thinking this way. You can recognize smart design. By becoming familiar with some common aspects and examples of great design, we will instill confidence in your visual instincts and learn some concrete tips to follow and adjustments to make when things don't feel quite right.

Affordances

In the field of design, experts speak of objects having "affordances." These are aspects inherent to the design that make it obvious how the product is to be used. For example, a knob affords turning, a button affords pushing, and a cord affords pulling. These characteristics suggest how the object is to be interacted with or operated. When sufficient affordances are present, good design fades into the background and you don't even notice it.

For an example of affordances in action, let's look to the OXO brand. On their website, they articulate their distinguishing feature as "Universal Design"—a philosophy of making products that are easy to use for the widest possible spectrum of users. Of particular relevance to our conversation here are their kitchen gadgets (which were once marketed as "tools you hold on to"). The gadgets are designed in such a way that there is really only one way to pick them up—the correct way. In this way, OXO kitchen gadgets afford correct use, without most users recognizing that this is due to thoughtful design (Figure 5.1).

FIGURE 5.1 OXO kitchen gadgets

Let's consider how we can translate the concept of affordances to communicating with data. We can leverage visual affordances to indicate to our audience how to use and interact with our visualizations. We'll discuss three specific lessons to this end: (1) highlight the important stuff, (2) eliminate distractions, and (3) create a clear hierarchy of information.

Highlight the important stuff

We've previously demonstrated the use of preattentive attributes to draw our audience's attention to where we want them to focus: in other words, to highlight the important stuff. Let's continue to explore this strategy. Critical here is to only highlight a fraction of the overall visual, since highlighting effects are diluted as the percentage that are highlighted increases. In *Universal Principles of Design* (Lidwell, Holden, and Butler, 2003), it is recommended that at most 10% of the visual design be highlighted. They offer the following guidelines:

- **Bold,** *italics,* and <u>underlining</u>: Use for titles, labels, captions, and short word sequences to differentiate elements. Bolding is generally preferred over italics and underlining because it adds minimal noise to the design while clearly highlighting chosen elements. Italics add minimal noise, but also don't stand out as much and are less legible. Underlining adds noise and compromises legibility, so should be used sparingly (if at all).
- CASE and typeface: Uppercase text in short word sequences is easily scanned, which can work well when applied to titles, labels, and keywords. Avoid using different fonts as a highlighting technique, as it's difficult to attain a noticeable difference without disrupting aesthetics.
- **Color** is an effective highlighting technique when used sparingly and generally in concert with other highlighting techniques (for example, bold).
- Inversing elements is effective at attracting attention, but can add considerable noise to a design so should be used sparingly.
- Size is another way to attract attention and signal importance.

I've omitted "blinking or flashing" from the list above, which Lidwell et al. include with instructions to use only to indicate highly critical information that requires immediate response. I do not recommend using blinking or flashing when communicating with data for explanatory purposes (it tends to be more annoying than helpful).

Note that preattentive attributes can be layered, so if you have something really important, you can signal this and draw attention by making it large, colored, and bold.

Let's look at a specific example using highlighting effectively in data visualization. A graph similar to Figure 5.2 was included in a February 2014 Pew Research Center article titled "New Census Data Show More Americans Are Tying the Knot, but Mostly It's the College-Educated."

New Marriage Rate by Education

Number of newly married adults per 1,000 marriage eligible adults

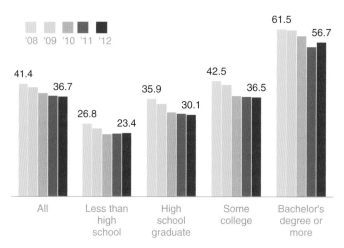

Note: Marriage eligible includes the newly married plus those widowed, divorced, or never married at interview.
Source: U.S. Census
Adapted from **PEW RESEARCH CENTER**

FIGURE 5.2   Pew Research Center original graph

Based on the article that accompanied it, Figure 5.2 is meant to demonstrate that the 2011 to 2012 increase observed in total new marriages was driven primarily by an increase in those having a bachelor's degree or more (there doesn't actually appear to be an increase based on the "All" trend shown, but let's ignore this). The design of Figure 5.2 does little to draw this clearly to our attention, however. Rather, my attention is drawn to the 2012 bars within the various groups because they are rendered in a darker color than the rest.

Changing the use of color in this visual can completely redirect our focus. See Figure 5.3.

New Marriage Rate by Education

Number of newly married adults per 1,000 marriage eligible adults

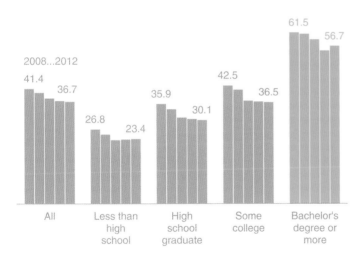

Note: Marriage eligible includes the newly married plus those widowed, divorced, or never married at interview.

Source: U.S. Census

Adapted from **PEW RESEARCH CENTER**

FIGURE 5.3   Highlight the important stuff

In Figure 5.3, the color orange has been used to highlight the data points for those having a bachelor's degree or more. By making everything else grey, the highlighting provides a clear signal of where we should focus our attention. We'll come back to this example momentarily.

Eliminate distractions

While we highlight the important pieces, we also want to eliminate distractions. In his book *Airman's Odyssey*, Antoine de Saint-Exupery famously said, "You know you've achieved perfection, not when you have nothing more to add, but when you have nothing to take away" (Saint-Exupery, 1943). When it comes to the perfection of design with data visualization, the decision of what to cut or de-emphasize can be even more important than what to include or highlight.

To identify distractions, think about both clutter and context. We've discussed clutter previously: these are elements that take up space but don't add information to our visuals. Context is what needs to be present for your audience in order for what you want to communicate to make sense. When it comes to context, use the right amount—not too much, not too little. Consider broadly what information is critical and what is not. Identify unnecessary, extraneous, or irrelevant items or information. Determine whether there are things that might be distracting from your main message or point. All of these are candidates for elimination.

Here are some specific considerations to help you identify potential distractions:

- **Not all data are equally important.** Use your space and audience's attention wisely by getting rid of noncritical data or components.
- **When detail isn't needed, summarize.** You should be familiar with the detail, but that doesn't mean your audience needs to be. Consider whether summarizing is appropriate.
- **Ask yourself: would eliminating this change anything?** No? Take it out! Resist the temptation to keep things because they are cute or because you worked hard to create them; if they don't support the message, they don't serve the purpose of communication.

- **Push necessary, but non-message-impacting items to the background**. Use your knowledge of preattentive attributes to de-emphasize. Light grey works well for this.

Each step in reduction and de-emphasis causes what remains to stand out more. In cases where you are unsure whether you'll need the detail that you're considering cutting, think about whether there is a way to include it without diluting your main message. For example, in a slide presentation, you can push content to the appendix so it's there if you need it but won't distract from your main point.

Let's look back at the Pew Research example discussed previously. In Figure 5.3, we used color sparingly to highlight the important part of our visual. We can further improve this graph by eliminating distractions, as illustrated in Figure 5.4.

New marriage rate by education
Number of newly married adults per 1,000 marriage eligible adults

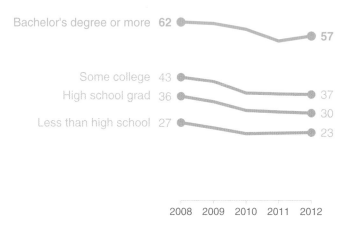

Note: Marriage eligible includes the newly married plus those widowed, divorced, or never married at interview.
Source: U.S. Census
Adapted from PEW RESEARCH CENTER

FIGURE 5.4   Eliminate distractions

In Figure 5.4, a number of changes were made to eliminate distractions. The biggest shift was from a bar graph to a line graph. As we've discussed, line graphs typically make it easier to see trends over time. This shift also has the effect of visually reducing discrete elements, because the data that was previously five bars has been reduced to a single line with the end points highlighted. When we consider the full data being plotted, we've gone from 25 bars to 4 lines. The organization of the data as a line graph allows the use of a single x-axis that can be leveraged across all of the categories. This simplifies the processing of the information (rather than seeing the years in a legend at the left and then having to translate across the various groups of bars).

The "All" category included in the original graph was removed altogether. This was the aggregate of all of the other categories, so showing it separately was redundant without adding value. This won't always be the case, but here it didn't add anything interesting to the story.

The decimal points in the data labels were eliminated by rounding to the nearest whole digit. The data being plotted is "Number of newly married adults per 1,000," and I find it strange to discuss the number of adults using decimal places (fractions of a person!). Additionally, the sheer size of the numbers and visible differences between them mean that we don't need the level of precision or granularity that decimal points provide. It is important to take context into account when making decisions like this.

The italics in the subtitle were changed to regular font. There was no reason to draw attention to these words. In the original, I found that the spatial separation between the title and subtitle also caused undue attention to be placed on the subtitle, so I removed the spacing in the makeover.

Finally, the highlighting of the "Bachelor's degree or more" category introduced in Figure 5.3 was preserved and extended to include the category name in addition to the data labels. As we've seen previously, this is a way to tie components together visually for our audience, easing the interpretation.

Figure 5.5 shows the before-and-after.

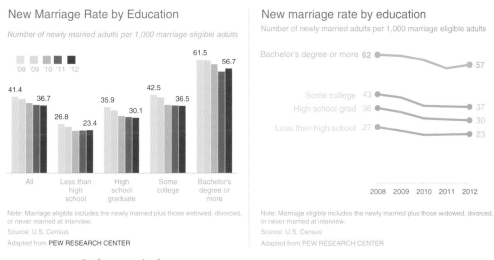

FIGURE 5.5   Before-and-after

By highlighting the important stuff and eliminating distractions, we've markedly improved this visual.

Create a clear visual hierarchy of information

As we discussed in Chapter 4, the same preattentive attributes we use to highlight the important stuff can be leveraged to create a hierarchy of information. We can visually pull some items to the fore-front and push other elements to the background, indicating to our audience the general order in which they should process the information we are communicating.

The power of super-categories

In tables and graphs, it can sometimes be useful to leverage super-categories to organize the data and help provide a construct for your audience to use to interpret it. For example, if you're looking at a table or graph that shows a value for 20 different demographic breakdowns, it can be useful to organize and clearly label the demographic breakdowns into groups or super-categories like age, race, income level, and education. These super-categories provide a hierarchical organization that simplifies the process of taking in the information.

Let's look at an example where a clear visual hierarchy has been established and discuss the specific design choices that were made to create it. Imagine you are a car manufacturer. Two important dimensions by which you judge the success of a particular make and model are (1) customer satisfaction and (2) frequency of car issues. A scatterplot could be useful to visualize how the current year's models compare with the previous year's average along these two dimensions, as shown in Figure 5.6.

Issues vs. **Satisfaction** by Model

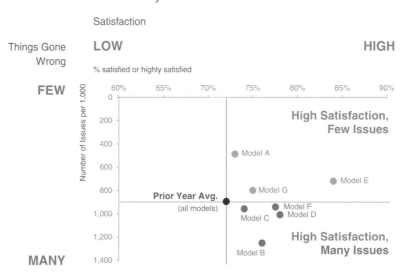

FIGURE 5.6   Clear visual hierarchy of information

Figure 5.6 lets us quickly see how this year's various models compare to last year's average on the basis of both satisfaction and issues. The size and color of font and data points alert us where to pay attention and in what general order. Let's consider the visual hierarchy of components and how they help us process the information presented. If I articulate the order in which I take in the information, it looks something like the following:

First, I read the graph title: "Issues vs. Satisfaction by Model." The bolding of Issues and Satisfaction signals that those words are important, so I have that context in mind as I process the rest of the visual.

Next, I see the *y*-axis primary label: "Things Gone Wrong." I note that these fall along a scale from few (at the top) to many (at the bottom). After that, I note the details across the horizontal *x*-axis: Satisfaction, ranging from low (left) to high (right).

I am then drawn to the dark grey point and corresponding words "Prior Year Average." The lines drawing this point to the axes allow me quickly to see that the prior year's average was around 900 issues per 1,000 and 72% satisfied or highly satisfied. This provides a useful construct for interpreting this year's models.

Finally, I am drawn to all of the red in the bottom right quadrant. The words tell me satisfaction is high, but there are many issues. It's clear because of how the visual is constructed that these are cases where the level of issues is greater than it was for last year's average. The red color reinforces that this is a problem.

We previously discussed super-categories for easing interpretation. Here, the quadrant labels "High Satisfaction, Few Issues" and "High Satisfaction, Many Issues" function in this manner. In absence of these, I could spend time processing the axis titles and labels and eventually figure out that's what these quadrants represent, but it's a much easier process when the pithy titles are present, eliminating the need for this processing altogether. Note that the left quadrants aren't labeled; labels are unnecessary since no values fall there.

Additional data points and details are there for context, but they are pushed to the background to reduce the cognitive burden and simplify the visual.

Upon sharing this visual with my husband, his reaction was "that's not the order I paid attention—I went straight to the red." That got me to thinking. First, I was surprised he started there, given that he's red-green colorblind, but he said that the red was different enough from everything else in the visual that it still grabbed his attention. Second, I look at so many graphs that it's ingrained in me to start with the details: the titles and axis titles to understand what I'm looking at before I get to the data. Others may look more quickly for the "so what." If we approach that way, we're drawn first to the bottom right quadrant since the red signals importance and that attention should be paid. After taking that in, perhaps we back up and read some of the other detail of the graph.

In either case, the thoughtful and clear visual hierarchy establishes order for the audience to use to process the information in a complex visual without it feeling, well, complicated. For our audience, by highlighting the important stuff, eliminating distractions, and establishing a visual hierarchy, the data visualizations we create afford understanding.

Accessibility

The concept of accessibility says that designs should be usable by people of diverse abilities. Originally, this consideration was for those with disabilities, but over time the concept has grown more general, which is the way in which I'll discuss it here. Applied to data visualization, I think of it as design that is usable by people of widely varying technical skills. You might be an engineer, but it shouldn't take someone with an engineering degree to understand your graph. As the designer, the onus is on *you* to make your graph accessible.

Poor design: who is at fault?

Well-designed data visualization—like a well-designed object—is easy to interpret and understand. When people have trouble understanding something, such as interpreting a graph, they tend to blame themselves. In most cases, however, this lack of understanding is not the user's fault; rather, it points to fault in the design. Good design takes planning and thought. Above all else, good design takes into account the needs of the user. This is another reminder to keep your user—your audience—top-of-mind when designing your communications with data.

For an example of accessibility in design, let's consider the iconic London underground tube map. Harry Beck produced a beautifully simple design in 1933, recognizing that the above-ground geography is unimportant when navigating the lines and removing the constraints it imposed. Compared to previous tube maps, Beck's accessible design rendered an easy-to-follow visual that became an essential guide to London and a template for transport maps around the world. It is that same map, with some minor modifications, that still serves London today.

We'll discuss two specific strategies related to accessibility in communicating with data: (1) don't overcomplicate and (2) text is your friend.

Don't overcomplicate

"If it's hard to read, it's hard to do." This was the finding of research undertaken by Song and Schwarz at the University of Michigan in 2008. First, they presented two groups of students with instructions for an exercise regimen. Half the students received the instructions written in easy-to-read Arial font; the other half were given instructions in a cursive-like font called Brushstroke. Students were asked how long the exercise routine would take and how likely they were to try it. The finding: the fussier the font, the more difficult the students judged the routine *and* the less likely they were to undertake it. A second study using a sushi recipe had similar findings.

Translation for data visualization: the more complicated it looks, the more time your audience perceives it will take to understand and *the less likely they are to spend time to understand it.*

As we've discussed, visual affordances can help in this area. Here are some additional tips to keep your visuals and communications from appearing overly complicated:

- **Make it legible:** use a consistent, easy-to-read font (consider both typeface and size).
- **Keep it clean:** make your data visualization approachable by leveraging visual affordances.
- **Use straightforward language:** choose simple language over complex, choose fewer words over more words, define any specialized language with which your audience may not be familiar, and spell out acronyms (at minimum, the first time you use them or in a footnote).
- **Remove unnecessary complexity:** when making a choice between simple and complicated, favor simple.

This is not about oversimplifying, but rather not making things more complicated than they need to be. I once sat through a presentation given by a well-respected PhD. The guy was obviously smart. When he said his first five-syllable word, I found myself impressed with his vocabulary. But as his academic language continued, I started to lose patience. His explanations were unnecessarily complicated. His words were unnecessarily long. It took a lot of energy to pay attention. I found it hard to listen to what he was saying as my annoyance grew.

Beyond annoying our audience by trying to sound smart, we run the risk of making our audience feel dumb. In either case, this is not a good user experience for our audience. Avoid this. If you find it hard to determine whether you are overcomplicating things, seek input or feedback from a friend or colleague.

Text is your friend

Thoughtful use of text helps ensure that your data visualization is accessible. Text plays a number of roles in communicating with data: use it to label, introduce, explain, reinforce, highlight, recommend, and tell a story.

There are a few types of text that absolutely must be present. Assume that every chart needs a title and every axis needs a title (exceptions to this rule will be *extremely* rare). The absence of these titles—no matter how clear you think it may be from context—causes your audience to stop and question what they are looking at. Instead, label explicitly so they can use their brainpower to understand the information, rather than spend it trying to figure out how to read the visual.

Don't assume that two different people looking at the same data visualization will draw the same conclusion. If there is a conclusion you want your audience to reach, state it in words. Leverage preattentive attributes to make those important words stand out.

Action titles on slides

The title bar at the top of your PowerPoint slide is precious real estate: use it wisely! This is the first thing your audience encounters on the page or screen and yet so often it gets used for redundant descriptive titles (for example, "2015 Budget"). Instead use this space for an action title. If you have a recommendation or something you want your audience to know or do, put it here (for example, "Estimated 2015 spending is above budget"). It means your audience won't miss it and also works to set expectations for what will follow on the rest of the page or screen.

When it comes to words in data visualization, it can sometimes be useful to annotate important or interesting points directly on a graph. You can use annotation to explain nuances in the data, highlight something to pay attention to, or describe relevant external factors. One of my favorite examples of annotation in data visualization is Figure 5.7 by David McCandless, "Peak Break-up Times According to Facebook Status Updates."

Peak Break-up Times
According to Facebook status updates

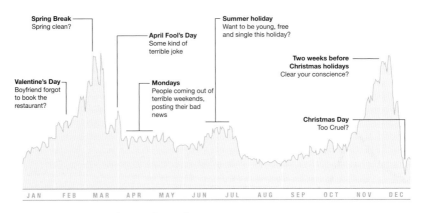

FIGURE 5.7   Words used wisely

As we follow the annotations from left to right in Figure 5.7, we see a small increase on Valentine's Day, then large peaks in the weeks of Spring Break (cleverly subtitled "Spring clean?"). There's a spike on April Fool's Day. The trend of break-ups on Mondays is highlighted. A gentle rise and fall in break-ups is observed over summer holiday. Then we see a massive increase leading up to the holidays, but a sharp drop-off at Christmas, because clearly breaking up with some-one then would simply be "Too Cruel."

Note how a few choice words and phrases make this data so much more quickly accessible than it otherwise would be.

As a side note, in Figure 5.7, the guidance I previously put forth about always titling the axes has not been followed. In this case, this is by design. Of more interest than the specific metric being plotted are the relative peaks and valleys. By not labeling the vertical axis (with either title or labels), you simply can't get caught up in a debate about it (What is being plotted? How is it being calculated? Do I agree with it?). This was a conscious design choice and won't be appropriate in most situations but, as we see in this case, can—in rare instances—work well.

In the context of accessibility via text, let's revisit the ticket example we examined in Chapters 3 and 4. Figure 5.8 shows where we left off after eliminating clutter and drawing attention to where we want our audience to focus via data markers and labels.

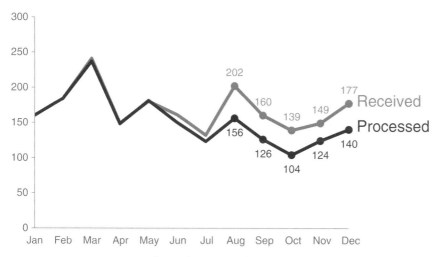

FIGURE 5.8   Let's revisit the ticket example

Figure 5.8 is a pretty picture, but it doesn't mean much without words to help us make sense of it. Figure 5.9 resolves this issue, adding the requisite text.

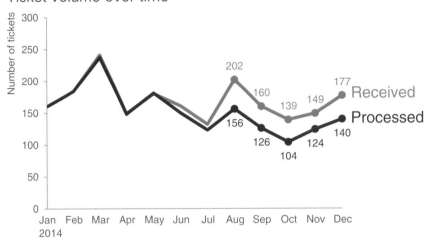

FIGURE 5.9   Use words to make the graph accessible

In Figure 5.9, we've added the words that have to be there: graph title, axis titles, and a footnote with the data source. In Figure 5.10, we take it a step further by adding a call to action and annotation.

Please approve the hire of 2 FTEs
to backfill those who quit in the past year

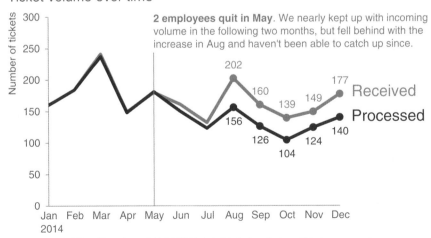

FIGURE 5.10   Add action title and annotation

In Figure 5.10, thoughtful use of text makes the design accessible. It's clear to the audience what they are looking at as well as what they should pay attention to and why.

Aesthetics

When it comes to communicating with data, is it really necessary to "make it pretty?" The answer is a resounding *Yes*. People perceive more aesthetic designs as easier to use than less aesthetic designs—whether they actually are or not. Studies have shown that more aesthetic designs are not only perceived as easier to use, but also more readily accepted and used over time, promote creative thinking and problem solving, and foster positive relationships, making people more tolerant of problems with designs.

A great example of the tolerance with problems that good aesthetics can foster is a former bottle design of Method liquid dishwashing soap, pictured in Figure 5.11. The anthropomorphic form rendered the soap an art piece—something to be displayed, not hidden away under the counter. This bottle design was wildly effective *in spite of* leakage issues. People were willing to overlook the inconvenience of the leaking bottle due to its appealing aesthetics.

FIGURE 5.11   Method liquid dishwashing soap

In data visualization—and communicating with data in general—spending time to make our designs aesthetically pleasing can mean our audience will have more patience with our visuals, increasing our chance of success for getting our message across.

If you aren't confident in your ability to create aesthetic design, look for examples of effective data visualization to follow. When you see a graph that looks nice, pause to consider what you like about it. Perhaps save it and build a collection of inspiring visuals. Mimic aspects from effective designs to create your own.

More specifically, let's discuss a few things to consider when it comes to aesthetic designs of data visualization. We've previously covered the main lessons relevant to aesthetics, so I'll touch on them here only briefly and then we'll discuss a specific example to see how being mindful of aesthetics can improve our data visualization.

1. **Be smart with color.** The use of color should always be an intentional decision; use color sparingly and strategically to highlight the important parts of your visual.

2. **Pay attention to alignment.** Organize elements on the page to create clean vertical and horizontal lines to establish a sense of unity and cohesion.

3. **Leverage white space.** Preserve margins; don't stretch your graphics to fill the space, or add things simply because you have extra space.

Thoughtful use of color, alignment, and white space are components of the design that you don't even notice when they are done well. But you notice when they aren't: rainbow colors, and lacking alignment and white space, make for a visual that's simply uncomfortable to look at. It feels disorganized and like no attention was paid to detail. This shows a lack of respect for your data and your audience.

Let's look at an example: see Figure 5.12. Imagine you work for a prominent U.S. retailer. The graph depicts the breakdown of the U.S. Population and Our Customers by seven customer segments (for example, age ranges).

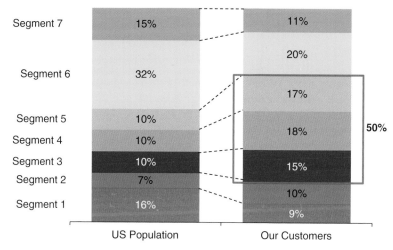

FIGURE 5.12   Unaesthetic design

We can leverage the lessons covered to make smarter design choices. Specifically, let's discuss how we can improve Figure 5.12 when it comes to the use of color, alignment, and white space.

Color is overused. There are too many colors, and they compete for our attention, making it difficult to focus on one at a time. Going back to the lesson on affordances, we should think about what we want to highlight to our audience and *only use color there*. In this case, the red box around segments 3 through 5 on the right signals that those segments are important, but there are so many things competing for our attention that it takes some time to even see that. We can make this a more obvious and easier process by using color strategically.

Elements are not properly aligned. The center alignment of the graph title makes it so it isn't aligned with anything else in the visual. The segment titles at the left aren't aligned to create a clean line either on the left or right. This looks sloppy.

Finally, white space is misused. There is too much of it between the segment titles and data, which makes it challenging to draw your eye from the segment title to the data (I have an urge to use my index finger to trace across: we can reduce white space between the titles and data, so this work is unnecessary). The white space between the columns of data is too narrow to optimally emphasize the data and cluttered with unneeded dotted lines.

Figure 5.13 shows how the same information could look if we remedy these design issues.

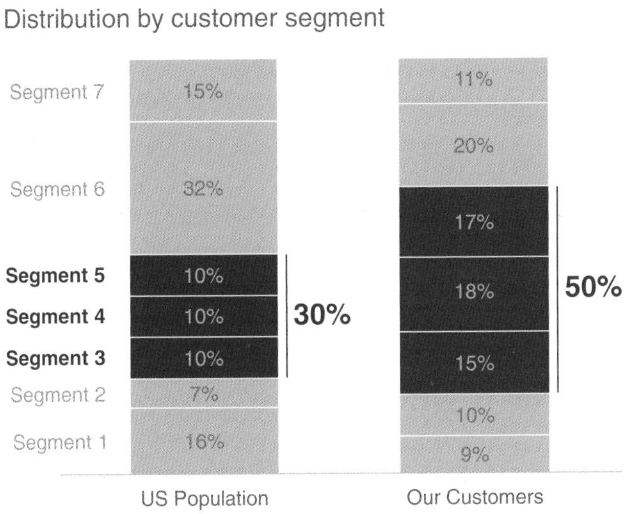

Distribution by customer segment

FIGURE 5.13   Aesthetic design

Aren't you more likely to spend a little more time with Figure 5.13? It is clear that attention to detail was paid to the design: it took the designer time to get this result. This creates a sort of onus on the part of the audience to spend time to understand it (this sort of contract doesn't exist with poor design). Being smart with color, aligning objects, and leveraging white space brings a sense of visual organization to your design. This attention to aesthetics shows a general respect for your work and your audience.

Acceptance

For a design to be effective, it must be accepted by its intended audience. This adage is true whether the design in question is that of a physical object or a data visualization. But what should you do when your audience isn't accepting of your design?

In my workshops, audience members regularly raise this dilemma: *I want to improve the way we look at things, but when I've tried to make changes in the past, my efforts have been met with resistance. People are used to seeing things a certain way and don't want us to mess with that.*

It is a fact of human nature that most people experience some level of discomfort with change. Lidwell et al. in *Universal Principles of Design* (2010) describe this tendency of general audiences to resist the new because of their familiarity with the old. Because of this, making significant changes to "the way we've always done it" may require more work to gain acceptance than simply replacing the old with the new.

There are a few strategies you can leverage for gaining acceptance in the design of your data visualization:

- **Articulate the benefits of the new or different approach.** Sometimes simply giving people transparency into *why* things will look different going forward can help them feel more comfortable. Are there new or improved observations you can make by looking at the data in a different way? Or other benefits you can articulate to help convince your audience to be open to the change?
- **Show the side-by-side.** If the new approach is clearly superior to the old, showing them side-by-side will demonstrate this. Couple this with the prior approach by showing the before-and-after *and* explaining why you want to shift the way you're looking at things.
- **Provide multiple options and seek input.** Rather than prescribing the design, consider creating several options and getting feedback

from colleagues or your audience (if appropriate) to determine which design will best meet the given needs.

- **Get a vocal member of your audience on board.** Identify influential members of your audience and talk to them one-on-one in an effort to gain acceptance of your design. Ask for their feedback and incorporate it. If you can get one or a couple of vocal members of your audience bought in, others may follow.

One thing to consider if you find yourself met with resistance is whether the root problem is that your audience is slow to change *or* if there might be issues with the design you are proposing. Test this by getting input from someone who doesn't have a vested interest. Show them your data visualization. If appropriate, also show the historical or current visuals. Have them talk you through their thought process as they review the visual. What do they like? What questions do they have? Which visual do they prefer and why? Hearing these things from a nonbiased third party may help you uncover issues with your design that are leading to the adoption challenge you face with your audience. The conversation may also help you articulate talking points that will help you drive the acceptance you seek from your audience.

In closing

By understanding and employing some traditional design concepts, we set ourselves up for success in communicating with data. Offer your audience visual affordances as cues for how to interact with your communication: highlight the important stuff, eliminate distractions, and create a visual hierarchy of information. Make your designs accessible by not overcomplicating and by leveraging text to label and explain. Increase your audience's tolerance of design issues by making your visuals aesthetically pleasing. Employ the strategies discussed for gaining audience acceptance for your visual designs.

Congratulations! You now know the 5th lesson in storytelling with data: how to **think like a designer**.

chapter six

dissecting model
visuals

Up to this point, we've covered a number of lessons you can employ to improve your ability to communicate with data. Now that you understand the basics of what makes a visual effective, let's consider some additional examples of what "good" data visualization looks like. Before covering our final lesson, in this chapter we will look at several model visuals and discuss the thought process and design choices that led to their creation, utilizing the lessons we've covered.

You'll notice some similar considerations being made across the various examples. When creating each example, I thought about how I want the audience to process the information and made corresponding choices regarding what to emphasize and draw the audience's attention to as well as what to de-emphasize. Because of this, you will see common points raised around color and size. The choice of visual, relative ordering of data, alignment and positioning of elements, and use of words are also discussed in a number of cases.

This repetition is useful to reinforce the concepts I'm thinking about and resulting design choices across the various examples.

Each visual highlighted was created to meet the need of a specific situation. I'll discuss the relevant scenarios briefly, but don't worry too much about the details. Rather, spend time looking at and thinking about each model visual. Consider what data visualization challenges you face where the given approach (or aspects of the given approach) could be leveraged.

Model visual #1: line graph

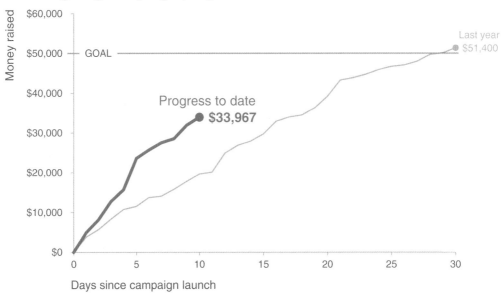

FIGURE 6.1   Line graph

Company X runs an annual month-long "giving campaign" to raise money for charitable causes. Figure 6.1 shows this year's progress to date. Let's consider what makes this example good and the deliberate choices made in the course of its creation.

Words are used appropriately. Everything is titled and labeled, so there's no question about what we are looking at. Graph title, vertical axis title, and horizontal axis title are present. The various lines in the graph are labeled directly, so there's no work going back and forth between a legend and the data to decipher what is being graphed. Good use of text makes this visual accessible.

If we apply the "where are your eyes drawn?" test described in Chapter 4, I briefly scan the graph title, then I'm drawn to the "Progress to date" trend (where we want the audience to focus). I almost always use dark grey for the graph title. This ensures that it stands out, but without the sharp contrast you get from pure black on white (rather, I preserve the use of black for a standout color when I'm not using any other colors). A number of preattentive attributes are employed to draw attention to the "Progress to date" trend: color, thickness of line, presence of data marker and label on the final point, and the size of the corresponding text.

When it comes to the broader context, a couple of points for comparison are included but de-emphasized so the graph doesn't become visually overwhelming. The goal of $50,000 is drawn on the graph for reference, but is pushed to the background by use of a thin line; both the line and text are the same grey as the rest of the graph details. Last year's giving over time is included but also de-emphasized through the use of a thinner line and lighter blue (to tie it visually with this year's progress, but without competing for attention).

A couple of deliberate decisions were made regarding axis labels. On the vertical y-axis, you could consider rounding the numbers to thousands—so the axis would range from $0 to $60 and the axis title would be changed to "Money raised (thousands of dollars)." If the numbers were on the scale of millions, I probably would have done this. For me, however, thinking about numbers in the thousands isn't as intuitive, so rather than mess with the scale here, I preserved the zeros in the y-axis labels.

On the horizontal x-axis, we don't need every single day labeled since we're more interested in the overall trend, not what happened on a specific day. Because we have data through the 10th day of a 30-day month, I chose to label every 5th day on the x-axis (given that this is days we're talking about, another potential solution would be to label every 7th day and/or add super-categories of week 1, week 2, etc.). This is one of those cases where there isn't a single right answer: you should think about the context, the data, and how you want your audience to use the visual and make a deliberate decision in light of those things.

Model visual #2: annotated line graph with forecast

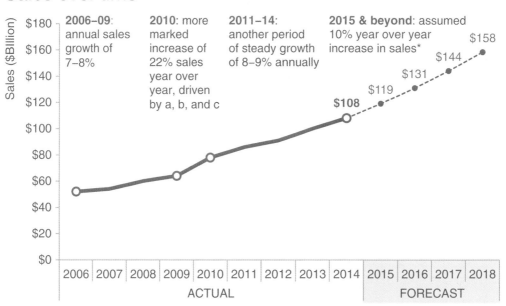

Data source: Sales Dashboard; annual figures are as of 12/31 of the given year.

*Use this footnote to explain what is driving the 10% annual growth forecast assumption.

FIGURE 6.2 Annotated line graph with forecast

Figure 6.2 shows an annotated line graph of actual and forecast annual sales.

I often see forecast and actual data plotted together as a single line, without any distinguishing aspects to set the forecast numbers apart from the rest. This is a mistake. We can leverage visual cues to draw a distinction between the actual and forecast data, easing the interpretation of the information. In Figure 6.2, the solid line represents actual data and a thinner dotted line (which carries some connotation of less certainty than a solid, bold line) represents the forecast data. Clear labeling of Actual and Forecast under the x-axis helps reinforce this (written in all caps for easy scanning), with the forecast portion set apart visually ever so slightly via light background shading.

In this visual, everything has been pushed to the background through the use of grey font and elements *except* the graph title, dates within the text boxes, data (line), select data markers, and numeric data labels from 2014 forward. When we consider the visual hierarchy of elements, my eye goes first to the graph title at the top left (due to both position and the preattentive larger dark grey text discussed in the prior example), then to the blue dates in the text boxes, at which point I can pause and read for a little context before moving my eye downward to see the corresponding point or trend in the data. Data markers are included only for those points referenced in the annotation, making it a quick process to see what part of the data is relevant to which annotation. (Originally, the data markers were solid blue, but I changed to white with blue outline, which made them stand out a little more in a way that I liked; the forecast data markers are smaller and solid blue, because white with blue outline there looked overly cluttered against the dotted lines.)

The $108 numeric label is bold. This is emphasized intentionally, since it is the last point of actual data and the anchor for the forecast. Historical data points are not labeled. Instead, the y-axis is preserved to give a general sense of magnitude, since we want the audience to focus on relative trends rather than precise values. Numeric data labels *are* included for the forecast data points to give the audience a clear understanding of forward-looking expectations.

All text in the visual is the same size except where intentional decisions were made to change it. The graph title is larger. The footnote is de-emphasized via smaller font and a low-priority placement at the bottom of the visual so that it is there to aid interpretation as needed, but doesn't draw attention.

Model visual #3: 100% stacked bars

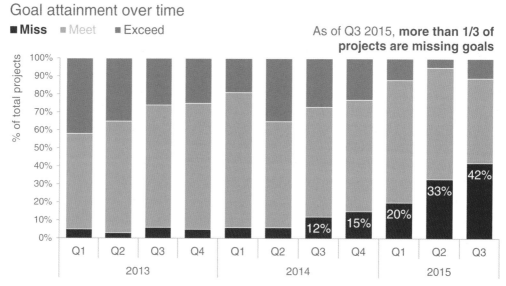

FIGURE 6.3   100% stacked bars

The stacked bar chart in Figure 6.3 is an example visual from the consulting world. Each consulting project has specific goals associated with it. Progress against those goals is assessed quarterly and designated as "Miss," "Meet," or "Exceed." The stacked bar chart shows the percentage of total projects in each of these categories over time. As with prior examples, don't worry too much about the details here; instead, reflect on what can be learned from the design considerations that went into creating this data visualization.

Let's first consider the alignment of objects within this visual. The graph title, legend, and vertical y-axis title are all aligned in the upper-left-most position. This means our audience encounters how to read the graph before they get to the data. On the left-hand side, the graph title, legend, y-axis title, and footnote are all aligned, creating a clean line on the left side of the visual. On the right-hand side, the text at the top is right-justified and aligned with the final bar of data that contains the data point being described (leveraging the Gestalt principle of proximity). This same text box is aligned vertically with the graph legend.

When it comes to focusing the audience's attention, red is used as the single attention-grabbing color (primary red tends to be too loud for me, so I often opt instead for a burnt-red shade as I did here). Everything else is grey. Numeric data labels were used—an additional visual cue signaling importance given the stark contrast of white on red and large text—on the points we want the audience to focus: the increasing percentage of projects missing goals. The rest of the data is preserved for context, but pushed to the background so it doesn't compete for attention. Slightly different shades of grey were used so you can still focus on one or the other series of data at a time, but it doesn't distract from the clear emphasis on the red series.

The categories fall along a scale from "Miss" to "Exceed," and this ordering is leveraged from bottom to top within the stacked bars. The "Miss" category is closest to the x-axis, making change over time easy to see because of the alignment of the bars at the same starting point (the x-axis). Change over time in the "Exceed" category is also easy because of the consistent alignment along the top of the graph. The change over time in the percentage of projects that meet their goals is harder to see because there is no consistent baseline at either the top or bottom of the graph, but given that this is a lower-priority comparison, this is OK.

Words make the visual accessible. The graph has a title, the y-axis has a title, and the x-axis leverages super-categories (years) to reduce redundant labeling and make the data more easily scannable. The words at the top right reinforce what we should be paying attention to (we will talk about words much more in the context of storytelling in Chapter 7). The footnote contains a note about the total number of projects over time, which is helpful context that we don't get from the visual directly due to the use of 100% stacked bars.

Model visual #4: leveraging positive and negative stacked bars

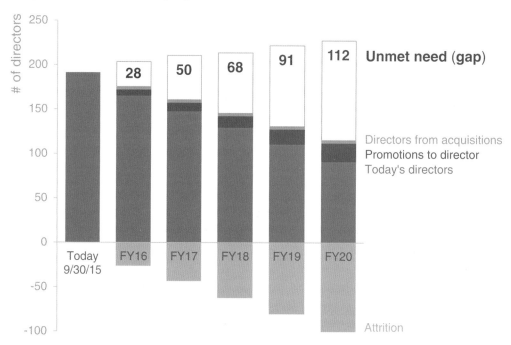

A footnote explaining relevant forecast assumptions and methodology would go here.

FIGURE 6.4 Leveraging positive and negative stacked bars

Figure 6.4 shows an example from the people analytics space. It can be useful to look forward to understand expected needs for senior talent and identify any gaps so they can be proactively addressed. In this example, there will be increasing unmet need for directors given assumptions for expected additions to the director pool over time through acquisitions and promotions and the decrease to the pool over time due to attrition (directors leaving the company).

If we consider the path our eyes take with Figure 6.4, mine scan the title, then go directly to the big, bold, black numbers and follow them to the right to the text that tells me this represents "Unmet need (gap)." My eye then goes downward, reading the text and glancing back leftward to the data each describes, until I hit the final series, "Attrition," at the bottom. At this point, my eyes sort of bounce back and forth between "Attrition" and "Unmet need (gap)" portions of the bars, noting that there is some increase in the total number of directors over time as we look left to right (likely as the overall company grows and the need for senior leaders increases as a result), but that the majority of the unmet need is due to attrition of the current director pool.

Intentional choices were made when it comes to the use of color throughout this visual. "Today's directors" are shown in my standard medium blue. The exiting directors ("Attrition") are shown in a less saturated version of the same color to tie these together visually. Over time, you see less of the blue falling above the axis and an increasing proportion falling below the axis as more and more directors attrite. The negative direction of the "Attrition" series reinforces that this volume represents a decrease to the director pool. Directors added through acquisitions and promotions are shown in green (which carries positive connotation). The unmet need is depicted by an outline only, to visually show empty space, reinforcing that this represents a gap. The text labels on the right are each written in the same color as the given data series they describe, except "Unmet need (gap)," which is written in the same big, bold, black text as the data labels for this series.

The ordering of the various data series within the stacked bars is deliberate. "Today's directors" is the base, and as such is shown beginning at the horizontal axis. As I mentioned previously, the negative "Attrition" series falls below that in a negative direction. Above "Today's directors" are the additions: promotions and acquisitions. Finally, at the top (where our eye hits sooner than the subsequent data), we encounter the "Unmet need (gap)."

The y-axis is preserved so the reader has a sense of total magnitude (both in the positive and negative direction), but it is pushed to the background via grey text. Only those specific points we should pay attention to—the "Unmet need (gap)"—are labeled directly with numerical values.

All text in the visual is the same size *except* where decisions were made to further emphasize or de-emphasize components. The graph title is larger. The axis title "# of directors" is slightly larger to ease the reading of the rotated text. The "Unmet need (gap)" text and numbers are bigger and bolder than anything else in the visual, as this is where we want the reader to pay attention. The footnote is written in smaller text, so it is there as needed but does not draw attention. By making it grey and in the lowest-priority position at the bottom of the visual, we further de-emphasize the footnote.

Model visual #5: horizontal stacked bars

Top 15 development priorities, according to survey

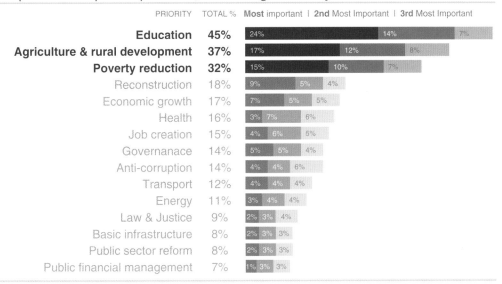

PRIORITY	TOTAL %	Most important	2nd Most Important	3rd Most Important
Education	**45%**	24%	14%	7%
Agriculture & rural development	**37%**	17%	12%	8%
Poverty reduction	**32%**	15%	10%	7%
Reconstruction	18%	9%	5%	4%
Economic growth	17%	7%	5%	5%
Health	16%	3%	7%	6%
Job creation	15%	4%	6%	5%
Governanace	14%	5%	5%	4%
Anti-corruption	14%	4%	4%	6%
Transport	12%	4%	4%	4%
Energy	11%	3%	4%	4%
Law & Justice	9%	2%	3%	4%
Basic infrastructure	8%	2%	3%	3%
Public sector reform	8%	2%	3%	3%
Public financial management	7%	1%	3%	3%

N = 4,392. Based on responses to item, *When considering development priorities, which one development priority is the most important? Which one is the second most important priority? Which one is the third most important priority?* Respondents chose from a list. Top 15 shown.

FIGURE 6.5   Horizontal stacked bars

Figure 6.5 shows the results of survey questions on relative priorities in a developing nation. This is a great deal of information, but due to strategic emphasizing and de-emphasizing of components, it does not become visually overwhelming.

Stacked bars make sense here given the nature of what is being graphed: top priority (in first position in the darkest shade), 2nd priority (in second position and a slightly lighter shade of the same color), and 3rd priority (in third position and an even lighter shade of the same color). Orienting the chart horizontally means the category names along the left are easy to read in horizontal text.

The categories are organized vertically in descending order of "Total %," giving the audience a clear construct to use as they interpret the data. The biggest categories are at the top, so we see them first. The top three priorities are emphasized specifically through the use of color (the narrative that accompanied the original version of this visual focused on these). This color is leveraged for the category name, total % and stacked bars of data. This consistent color ties the components together visually.

One decision point when graphing data is whether to preserve the axis, label the data points (or some data points) directly, or both. In this case, the numeric data labels within the bars have been pre-served, but de-emphasized with smaller text (oriented to the left, which creates a clean line as you scan down the data labels for the "Most important," making it feel slightly less cluttered than right- or center-oriented text that would vary in position across each of the bars). The data labels were further de-emphasized through the color they are written in: a light shade of blue or grey that doesn't create as stark a contrast as white labels on a colored bar. The x-axis was eliminated altogether. Here, we implicitly assume that the specific values are important enough to label. Another scenario may call for a different approach.

As we noted with a number of the previous examples, words are used well in this visual. Everything is titled and labeled. The titles "Prior-ity" and "Total %" are written all in caps for easy scanning. The leg-end for the interpretation of the bars appears immediately above the first bar of data with the keywords "Most," "2nd," and "3rd" bolded for emphasis. Additional detail is described in the footnote.

In closing

We can learn by examining effective visual displays and consider-ing the design choices that were made to create them. Through the examples in this chapter, we've reinforced a number of the les-sons covered up to this point. We touched on the choice of graph type and ordering of data. We considered where our eyes are drawn and in what order due to strategies employed to emphasize and de-emphasize components through the use of color, thickness, and size. We discussed the alignment and positioning of elements. We considered the appropriate use of text that makes the visuals accessible through clear titling, labeling, and annotation.

There is something to be learned from every example of data visualization you encounter—both good and bad. When you see something you like, pause to consider *why*. Those who follow my blog (storytellingwithdata.com) might be aware that I'm also an avid cook, and I often parlay the following food metaphor into data analytics: in data visualization, there is rarely (if ever) a single "right" answer; rather, there are flavors of good. The examples we've looked at in this chapter are the haute cuisine of charts.

That said, different people will make different decisions when faced with the same data visualization challenge. Because of this, inevitably I've made some design choices in these visuals that you might have handled differently. That's OK. I hope by articulating my thought process that you can understand why I made the design choices I did. These are considerations to keep in mind in your own design process. Of primary importance is that your design choices be just that: intentional.

Now you're ready for the final storytelling with data lesson: **tell a story**.

chapter seven

lessons in storytelling

In my workshops, the lesson on storytelling often begins with a thought exercise. I ask participants to close their eyes and recall the story of *Red Riding Hood*, considering specifically the plot, the twists, and the ending. This exercise sometimes generates some laughs; people wonder about its relevance or gamely confuse it with *Three Little Pigs*. But I find that the majority of participants (typically around 80–90% based on a show of hands) are able to remember the high-level story—often a modified version of Grimms' macabre original.

Indulge me for a moment, while I tell you the version that resides in my head:

Grandma has fallen ill and Red Riding Hood sets out on a walk through the woods with a basket of goodies to deliver to her. On her way, she encounters a woodsman and a wolf. The wolf runs ahead, eats Grandma, and dresses up in her clothes. When Red arrives, she senses something is awry. She goes through a series of questions with the wolf (posing as Grandma), culminating in the observation:

"Oh, Grandma, how big your teeth are!"—to which the wolf replies, "The better to eat you with!" and swallows Red whole. The woodsman walks by, and, seeing the door to Grandma's house ajar, decides to investigate. Inside, he finds the wolf dozing after his meal. The woodsman suspects what has happened and chops the wolf in half. Grandma and Red Riding Hood emerge—safe and sound! It is a happy ending for everyone (except the wolf).

Now let's turn back to the question that may be on the tip of your tongue: What could *Red Riding Hood* possibly have to do with communicating with data?

For me, this exercise is evidence of a couple of things. First is the power of repetition. You likely have heard some version of *Red Riding Hood* a number of times. Perhaps you've read or told a version of the story a number of times. This process of hearing, reading, and saying things numerous times helps to cement them in our long-term memory. Second, stories like *Red Riding Hood* employ this magical combination of plot-twists-ending (or, as we'll learn momentarily from Aristotle—beginning, middle, and end), which works to embed things in our memory in a way that we can later recall *and retell* the story to someone else.

In this chapter, we explore the magic of **story** and how we can use concepts of storytelling to communicate effectively with data.

The magic of story

When you see a great play, watch a captivating movie, or read a fantastic book, you've experienced the magic of story. A good story grabs your attention and takes you on a journey, evoking an emotional response. In the middle of it, you find yourself not wanting to turn away or put it down. After finishing it—a day, a week, or even a month later—you could easily describe it to a friend.

Wouldn't it be great if we could ignite such energy and emotion in our audiences? Story is a time-tested structure; humans have been communicating with stories throughout history. We can leverage this powerful tool for our business communications. Let's look to the art forms of plays, movies, and books to understand what we can learn from master storytellers that will help us better tell our own stories with data.

Storytelling in plays

The notion of narrative structure was first described in ancient times by Greek philosophers such as Aristotle and Plato. Aristotle introduced a basic but profound idea: that story has a clear beginning, middle, and end. He proposed a three-act structure for plays. This concept has been refined over time and is commonly referred to as the setup, conflict, and resolution. Let's look briefly at each of these acts and what they contain, and then we'll consider what we can learn from this approach.

The first act sets up the story. It introduces the main character, or protagonist, their relationships, and the world in which they live. After this setup, the main character is confronted with an incident. The attempt to deal with this incident typically leads to a more dramatic situation. This is known as the first turning point. The first turning point ensures that life will never be the same for the main character and raises the dramatic question—framed in terms of the main character's call to action—to be answered in the climax of the play. This marks the end of the first act.

The second act makes up the bulk of the story. It depicts the main character's attempt to resolve the problem created through the first turning point. Often, the main character lacks the skills to deal with the problem he faces, and, as a result, finds himself encountering increasingly worsening situations. This is known as the character arc, where the main character goes through major changes in his life as a result of what is happening. He may have to learn new skills or reach a higher sense of awareness of who he is and what he is capable of in order to deal with his situation.

The third act resolves the story and its subplots. It includes a climax, where the tensions of the story reach the highest point of intensity. Finally, the dramatic question introduced in the first act is answered, leaving the protagonist and other characters with a new sense of who they really are.

There are a couple of lessons to be learned here. First, the three-act structure can serve as a model for us when it comes to communicating in general. Second, that **conflict** and **tension** are an integral part of story. We'll come back to these ideas shortly and explore some concrete applications. In the meantime, let's see what we can learn from an expert storyteller from the movies.

Storytelling and the cinema

Robert McKee is an award-winning writer and director and a well-respected screenwriting lecturer (his former students include 63 Academy Award and 164 Emmy Award winners, and his book, *Story*, is required reading in many university cinema and film programs). In an interview for *Harvard Business Review*, he discusses persuasion through storytelling and examines how storytelling can be leveraged in a business setting. McKee says there are two ways to persuade people:

The first is conventional rhetoric. In the business world, this typically takes the form of PowerPoint slides filled with bulleted facts and statistics. It's an intellectual process. But it is problematic, because while you're trying to persuade your audience, they are arguing with you in their heads. McKee says, "If you do succeed in persuading them, you've only done so on an intellectual basis. That's not good enough, because people are not inspired to act by reason alone" (Fryer, 2003).

Think about what *Red Riding Hood* would look like if we reduced the story to conventional rhetoric. Libby Spears does an amusing version of this in her slide deck, *Little Red Riding Hood and the Day PowerPoint Came to Town*. Here is my take on it—bullets on a Power-Point slide might look something like the following:

- Red Riding Hood (RRH) has to walk 0.54 mi from Point A (home) to Point B (Grandma's)

- RRH meets Wolf, who (1) runs ahead to Grandma's, (2) eats her, and (3) dresses in her clothes
- RRH arrives at Grandma's at 2PM, asks her three questions
- Identified problem: after third question, Wolf eats RRH
- Solution: vendor (Woodsman) employs tool (ax)
- Expected outcome: Grandma and RRH alive, wolf is not

When reduced to the facts, it's not so interesting, is it?

The second way to persuade, according to McKee, is through *story*. Stories unite an idea with an emotion, arousing the audience's attention and energy. Because it requires creativity, telling a compelling story is harder than conventional rhetoric. But delving into your creative recesses is worth it because story allows you to engage your audience on an entirely new level.

What exactly is *story*? At a fundamental level, a story expresses how and why life changes. Stories start with balance. Then something happens—an event that throws things out of balance. McKee describes this as "subjective expectation meets cruel reality." This is that same tension we discussed in the context of plays. The resulting struggle, conflict, and suspense are critical components of the story.

McKee goes on to say that stories can be revealed by asking a few key questions: *What does my protagonist want in order to restore balance in his or her life? What is the core need? What is keeping my protagonist from achieving his or her desire? How would my protagonist decide to act in order to achieve his or her desire in the face of those antagonistic forces?* After creating the story, McKee suggests leaning back to consider: *Do I believe this? Is it neither an exaggeration nor a soft-soaping of the struggle? Is this an honest telling, though heaven may fall?*

What can we learn from McKee? The meta-lesson is that we can use stories to engage our audience emotionally in a way that goes beyond what facts can do. More specifically, we can use the questions he outlines to identify stories to frame our communications. We'll consider this further soon. First, let's see what we can learn about storytelling from a master storyteller when it comes to the written word.

Storytelling and the written word

When asked about writing a captivating story by *International Paper*, Kurt Vonnegut (author of novels such as *Slaughterhouse-Five* and *Breakfast of Champions*) outlined the following tips, which I've excerpted from his short article, "How to Write with Style" (a great quick read):

1. **Find a subject you care about.** It is this genuine caring, and not your games with language, which will be the most compelling and seductive element in your style.

2. **Do not ramble, though.**

3. **Keep it simple.** Great masters wrote sentences which were almost childlike when their subjects were most profound. "To be or not to be?" asks Shakespeare's Hamlet. The longest word is three letters.

4. **Have the guts to cut.** If a sentence, no matter how excellent, does not illuminate your subject in some new and useful way, scratch it out.

5. **Sound like yourself.** I myself find that I trust my own writing most, and others seem to trust it most, too, when I sound most like a person from Indianapolis, which is what I am.

6. **Say what you meant to say.** If I broke all the rules of punctuation, had words mean whatever I wanted them to mean, and strung them together higgledy-piggledy, I would simply not be understood.

7. **Pity the readers.** Our audience requires us to be sympathetic and patient teachers, ever willing to simplify and clarify.

This advice contains a number of gems that we can apply in the context of storytelling. Keep it simple. Edit ruthlessly. Be authentic.

Don't communicate for yourself—communicate *for your audience.* The story is not for you; the story is for them.

Now that we've learned some lessons from the masters, let's consider how we can construct our stories.

Constructing the story

We introduced the foundations of a narrative in Chapter 1 with the Big Idea, 3-minute story, and storyboarding to outline content to include while starting to consider order and flow. We learned how important it is to identify our audience—both who they are and what we need them to do. In the interim, we also learned how to perfect the data visualizations that we'll include in our communication. Now that we're set on that front, it is time to turn back to the story. Story is what ties together information, giving our presentation or communication a framework for our audience to follow.

Perhaps Vonnegut appreciated Aristotle's simple yet profound observation that a story has a clear beginning, middle, and end. For a concrete example, think back to what we considered with *Red Riding Hood:* the magical combination of plot, twists, and ending. We can use this idea of beginning, middle, and end—taking inspiration from the three-act structure—to set up the stories that we want to communicate with data. Let's discuss each of these pieces and the specifics to consider when crafting your story.

The beginning

The first thing to do is introduce the **plot**, building the context for your audience. Consider this the first act. In this section, we set up the essential elements of story—the setting, main character, unresolved state of affairs, and desired outcome—getting everyone on common ground so the story can proceed. We should involve our audience, piquing their interest and answering the questions that are likely on their mind: *Why should I pay attention? What is in it for me?*

In his book, *Beyond Bullet Points*, Cliff Atkinson outlines the following questions to consider and address when it comes to setting up the story:

1. The setting: When and where does the story take place?

2. The main character: Who is driving the action? (This should be framed in terms of your audience!)

3. The imbalance: Why is it necessary, what has changed?

4. The balance: What do you want to see happen?

5. The solution: How will you bring about the changes?

Note the similarity between the questions above and those raised by McKee that we covered earlier.

Using PowerPoint to tell stories

Cliff Atkinson uses PowerPoint to tell stories, leveraging the basic architecture of the three-act structure. His book, *Beyond Bullet Points*, introduces a story template and offers practical advice using PowerPoint to help users create stories with their presentations. More on this and related resources can be found at beyondbulletpoints.com.

Another way to think about the imbalance-balance-solution in your communication is to frame it in terms of the problem and your recommended solution. If you find yourself thinking, *But I don't have a problem!*—you may want to reconsider. As we've discussed, conflict and dramatic tension are critical components of a story. A story where everything is rosy and is expected to continue to be is not so interesting, attention-grabbing, or action-inspiring. Think of conflict and tension—between the imbalance and balance, or in terms of the problem on which you are focusing—as the storytelling tools that will help you to engage your audience. Frame your story in terms of their (your audience's) problem so that they immediately have a stake in the solution. Nancy Duarte calls this tension "the conflict between what *is* and what *could be*." There is always a story to tell. If it's worth communicating, it's worth spending the time necessary to frame your data in a story.

The middle

Once you've set the stage, so to speak, the bulk of your communication further develops "what could be," with the goal of convincing your audience of the need for action. You retain your audience's attention through this part of the story by addressing *how* they can solve the problem you introduced. You'll work to convince them *why* they should accept the solution you are proposing or act in the way you want them to.

The specific content will take different forms depending on your situation. The following are some ideas for content that might make sense to include as you build out your story and convince your audience to buy in:

- Further develop the situation or problem by covering relevant background.
- Incorporate external context or comparison points.
- Give examples that illustrate the issue.
- Include data that demonstrates the problem.
- Articulate what will happen if no action is taken or no change is made.
- Discuss potential options for addressing the problem.
- Illustrate the benefits of your recommended solution.
- Make it clear to your audience why they are in a unique position to make a decision or drive action.

When considering what to include in your communication, keep your audience top of mind. Think about what will resonate with them and motivate them. For example, will your audience be motivated to act by making money, beating the competition, gaining market share, saving a resource, eliminating excess, innovating, learning a skill, or something else? If you can identify what motivates your audience, consider framing your story and the need for action in terms of this. Also think about whether and when data will strengthen your story and integrate it as makes sense. Throughout your communication, make the information specific and relevant to your audience. The story should ultimately be about your audience, not about you.

Write the headlines first

When it comes to structuring the flow of your overall presentation or communication, one strategy is to create the headlines first. Think back to the storyboarding that we discussed in Chapter 1. Write each headline on a Post-it note. Play with the order to create a clear flow, connecting each idea to the next in a logical fashion. Establishing this sort of structure helps ensure that there is a logical order for your audience to follow. Make each headline the title of your presentation slides or the section title in a written report.

The end

Finally, the story must have an end. End with a **call to action**: make it totally clear to your audience what you want them to *do* with the new understanding or knowledge that you've imparted to them. One classic way to end a story is to tie it back to the beginning. At the beginning of our story, we set up the plot and introduced the dramatic tension. To wrap up, you can think about recapping this problem and the resulting need for action, reiterating any sense of urgency and sending your audience off ready to act.

When it comes to the order and telling of our story, another important consideration is the narrative structure, which we'll discuss next.

The narrative structure

In order to be successful, a narrative has to be central to the communication. These are words—written, spoken, or a combination of the two—that tell the story in an order that makes sense and convinces the audience why it's important or interesting and attention to it should be paid.

The most beautiful data visualization runs the risk of falling flat without a compelling narrative to go with it.

You've perhaps experienced this before if you've ever sat through a great presentation that used run-of-the-mill slides. A skilled presenter can overcome mediocre materials. A strong narrative can overcome less-than-ideal visuals. This is not to say that you shouldn't spend time making your data visualizations and visual communications great, but rather to underscore the importance of a compelling and robust narrative. Nirvana in communicating with data is reached when the effective visuals are combined with a powerful narrative.

Let's discuss some specific considerations when it comes to both the order of the story and the spoken and written narrative.

Narrative flow: the order of your story

Think about the order in which you want your audience to experience your story. Are they a busy audience who will appreciate if you lead with what you want from them? Or are they a new audience, with whom you need to establish credibility? Do they care about your process or just want the answer? Is it a collaborative process through which you need their input? Are you asking them to make a decision or take an action? How can you best convince them to act in the way you want them to? The answers to these questions will help you to determine what sort of narrative flow will work best, given your specific situation.

One important basic point here is that your story must have an order to it. A collection of numbers and words on a given topic without structure to organize them and give them meaning is useless. The narrative flow is the spoken and written path along which you take your audience over the course of your presentation or communication. This path should be clear to you. If it isn't, there certainly isn't a way to make it clear to your audience.

Help me turn this into a story!

When a client comes to me with a presentation deck and asks for help, the first thing I have them do is set the deck aside. I walk them through exercises that help them articulate the Big Idea and 3-minute story that we discussed in Chapter 1. Why? You have to have a solid understanding of what you want to communicate before you craft the communication. Once you have the Big Idea and 3-minute story articulated, you can start to think about what narrative flow makes sense and how to organize your deck.

One way to do this is to include a slide at the beginning of the deck that bullets the main points in your story. This will become an executive summary that says to your audience at the onset of the presentation, "here's what we will cover in our time together." Then organize the remaining slides to follow this same flow. Finally, at the end of the presentation, you'll repeat this ("here's what we covered") with emphasis on any actions you need your audience to take, or any decisions you need them to make. This helps to establish a structure to your presentation and make that structure clear to your audience. It also leverages the power of repetition to help your message stick with your audience.

One way to order the story—the one that typically comes most naturally—is **chronologically.** By way of example, if we think about the general analytical process, it looks something like this: we identify a problem, we gather data to better understand the situation, we analyze the data (look at it one way, look at it another way, tie in other things to see if they had an impact, etc.), we emerge with a finding or solution, and based on this we have a recommended action. One way to approach the communication of this to our audience is to follow that same path, taking the audience through it in the same way we experienced it. This approach can work well if you need to establish credibility with your audience, or if you know they care about the process. But chronological is not your only option.

Another strategy is to **lead with the ending**. Start with the call to action: what you need your audience to know or do. Then back up into the critical pieces of the story that support it. This approach can work well if you've already established trust with your audience or you know they are more interested in the "so what" and less interested in how you got there. Leading with the call to action has the additional benefit of making it immediately clear to your audience what role they are meant to play or what lens they should have on as they consider the rest of your presentation or communication, and why they should keep listening.

As part of making the narrative flow clear, we should consider what pieces of the story will be written and what will be conveyed through spoken words.

The spoken and written narrative

If you're giving a presentation—whether formally standing in front of a room, or more informally seated around a table—a good portion of the narrative will be spoken. If you're sending an email or report, the narrative is likely entirely written. Each format presents its own opportunities and challenges.

With a **live presentation**, you have the benefit of words on the screen or page being reinforced by the words you are saying. In this manner, your audience has the opportunity to both read and hear what they need to know, strengthening the information. You can use your voiceover to make the "so what" of each visual clear, make it relevant to your audience, and tie one idea to the next. You can respond to questions and clarify as needed. One challenge with a live presentation is that you must ensure what your audience needs to read on a given slide or section isn't so dense or consuming that their attention is focusing on that instead of listening to you.

Another challenge is that your audience can act unpredictably. They can ask questions that are off topic, jump to a point later in the presentation, or do other things to push you off track. This is one reason it's important—especially in a live presentation setting—to articulate clearly the role you want your audience to play and how your presentation is structured. For example, if you're anticipating an audience who will want to go off track, start by saying something like, "I know you are going to have a lot of questions. Write them down as they come up and I will make sure to leave time at the end to address any that aren't answered. But first, let's take a look at the process our team went through to reach our conclusion, which will lead us to what we are asking of you today."

As another example, if you're planning to lead with the ending and this differs from the typical approach—tell your audience that this is what you're doing. You might say something like, "Today, I'm going to start with what we're asking of you. The team did some robust analysis that led us to this conclusion and we weighed several different options. I will take you through all of this. But before I do, I want to spotlight what we are asking of you today, which is …" By telling your audience how you are going to structure your presentation, it can make both you and them more comfortable. It helps your audience to know what to expect and what role they are meant to play.

In a **written report** (or a presentation deck that is sent around instead of presented or also used as a "leave behind" to remind people of the content after you've delivered the presentation), you don't have the benefit of the voiceover to make the sections or slides relevant—rather, they must do this on their own. The written narrative is what will achieve this. Think about what words need to be present. In the case when something will be sent around without you there to explain it, it's especially important to make the "so what" of each slide or section clear. You've probably experienced when this has not been done well: you're looking through a presentation and encounter a slide of bulleted facts, or a graph or table packed with numbers, and are thinking, "I have no idea what I'm meant to get out of this." Don't let this happen to your work: make sure the words are present to make your point clear and relevant to your audience.

Getting feedback from someone not as familiar with the topic can be especially useful in this situation. Doing so will help you uncover issues with clarity and flow, or questions your audience may have, so you can address those proactively. In terms of benefits of the written report approach, if you make your structure clear, your audience can turn directly to the parts that interest them.

While we establish narrative structure and flow, the power of repetition is another strategy we can leverage within our storytelling.

The power of repetition

Thinking back to *Red Riding Hood*, one of the reasons I remember the story is due to repetition. I was told and read the story countless times as a little girl. As we discussed in Chapter 4, important information is gradually transferred from short-term memory into long-term memory. The more the information is repeated or used, the more likely it is to eventually end up in long-term memory, or to be retained. That's why the story of *Red Riding Hood* remains in my head today. We can leverage this power of repetition in the stories we tell.

Repeatable sound bites

"If people can easily recall, repeat, and transfer your message, you did a great job conveying it." To help facilitate this, Nancy Duarte recommends leveraging repeatable sound bites: succinct, clear, and repeatable phrases. Check out her book, *Resonate*, to learn more.

When it comes to employing the power of repetition, let's explore a concept called **Bing, Bang, Bongo**. My junior high English teacher introduced this idea to me when we were learning to write essays. The concept stuck with me—perhaps due to the consonance of the "Bing, Bang, Bongo" name and my teacher's use of it as a repeatable sound bite—and it can be leveraged when we need to tell a story with data.

The idea is that you should first tell your audience what you're going to tell them ("Bing," the introduction paragraph in your essay). Then you tell it to them ("Bang," the actual essay content). Then you summarize what you just told them ("Bongo," the conclusion). Applying this to a presentation or report, you can start with an executive summary that outlines for your audience what you are going to cover, then you can provide the detail or main content of your presentation, and finally end with a summary slide or section that reviews the main points you covered (Figure 7.1).

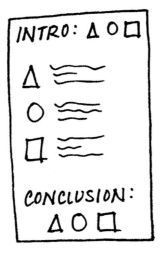

FIGURE 7.1   Bing, bang, bongo

If you're the one preparing or giving the presentation or writing the report, this may feel redundant, since you're already familiar with the content. But to your audience—who is not as close to the content—it feels nice. You've set their expectations on what you're going to cover, then provided detail, and then recapped. The repetition helps cement it in their memory. After hearing your message three times, they should be clear on what they are meant to know and do from the story you've just told.

Bing, Bang, Bongo is one strategy to leverage to help ensure that your story is clear. Let's consider some additional tactics.

Tactics to help ensure that your story is clear

There are a number of concepts I routinely discuss in my workshops for helping to ensure that the story you're telling in your communication comes across. These apply mainly to a presentation deck. While not always the case, I find that this is often the primary form of communicating analytical results, findings, and recommendations at many companies. Some of the concepts we'll discuss will be applicable to written reports and other formats as well.

Let's discuss four tactics to help ensure that your story is clear in your presentation: horizontal logic, vertical logic, reverse storyboarding, and a fresh perspective.

Horizontal logic

The idea behind horizontal logic is that you can read *just the slide title* of each slide throughout your deck and, together, these snippets tell the overarching story you want to communicate. It is important to have action titles (not descriptive titles) for this to work well.

One strategy is to have an executive summary slide up front, with each bullet corresponding to a subsequent slide title in the same order (Figure 7.2). This is a nice way of setting it up so your audience knows what to expect and then is taken through the detail (think back to the Bing, Bang, Bongo approach we covered previously).

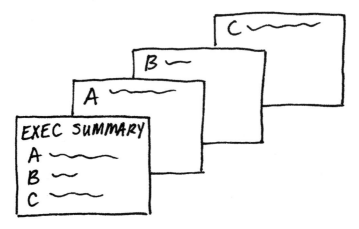

FIGURE 7.2   Horizontal logic

Checking for horizontal logic is one approach to test whether the story you want to tell is coming through clearly in your deck.

Vertical logic

Vertical logic means that all information on a given slide is self-reinforcing. The content reinforces the title and vice versa. The words reinforce the visual and vice versa (Figure 7.3). There isn't any extraneous or unrelated information. Much of the time, the decision on what to eliminate or push to an appendix is as important (sometimes more so) as the decision on what to retain.

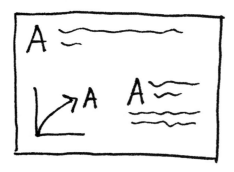

FIGURE 7.3   Vertical logic

Employing horizontal and vertical logic together will help ensure that the story you want to tell comes across clearly in your communication.

Reverse storyboarding

When you storyboard at the onset of building a communication, you craft the outline of the story you intend to tell. As the name implies, reverse storyboarding does the opposite. You take the final communication, flip through it, and write down the main point from each page (it's a nice way to test your horizontal logic as well). The resulting list should look like the storyboard or outline for the story you want to tell (Figure 7.4). If it doesn't, this can help you understand structurally where you might want to add, remove, or move pieces around to create the overall flow and structure for the story that you're interested in conveying.

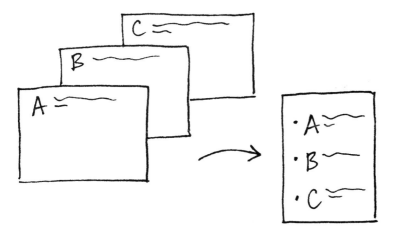

FIGURE 7.4   Reverse storyboarding

A fresh perspective

We've discussed the value of a fresh perspective to help see through your audience's lens when it comes to your data visualization (Figure 7.5). Seeking this sort of input for your overall presentation can be immensely helpful as well. Once you've crafted your communication, give it to a friend or colleague. It can be someone without any context (it's actually helpful if it *is* someone without any context, because this puts them in a much closer position to your audience than you can be, given your intimate knowledge of the subject matter). Ask them to tell you what they pay attention to, what they think is important, and where they have questions. This will help you understand whether the communication you've crafted is telling the story you mean to tell or, in the case where it isn't exactly, help you identify where to concentrate your iterations.

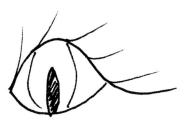

FIGURE 7.5   A fresh perspective

There is incredible value in getting a fresh perspective when it comes to communicating with data in general. As we become subject matter experts in our space, it becomes impossible for us to take a step back and look at what we've created (whether a single graph or a full presentation) through our audience's eyes. But that doesn't mean you can't see what they see. Leverage a friend or colleague for their fresh perspective. Help ensure that your communication hits the mark.

In closing

Stories are magical. They have the power of captivating us and sticking with us in ways that facts alone cannot. They lend structure. Why wouldn't you leverage this potential when crafting your communications?

When we construct stories, we should do so with a beginning (plot), middle (twists), and end (call to action). Conflict and tension are key to grabbing and maintaining your audience's attention. Another central component to story is the narrative, which we should consider in terms of both order (chronological or lead with ending) and manner (spoken, written, or a combination of the two). We can utilize the power of repetition to help our stories stick with our audience. Tactics such as horizontal and vertical logic, reverse storyboarding, and seeking a fresh perspective can be employed to help ensure that our stories come across clearly in our communications.

The main character in every story we tell should be the same: our audience. It is by making our audience the protagonist that we can ensure the story is about *them*, not about *us*. By making the data we want to show relevant to our audience, it becomes a pivotal point in our story. No longer will you just show data. Rather, you will tell a story with data.

With that, you can consider your final lesson learned. You now know how to **tell a story**.

Next, let's look at an example of the entire storytelling with data process, from start to finish.

chapter eight

pulling it all together

Up to this point, we've focused on individual lessons that, together, set you up for success when it comes to effectively visualizing and communicating with data. To refresh your memory, we've covered the following lessons:

1. Understand the context (Chapter 1)

2. Choose an appropriate display (Chapter 2)

3. Eliminate clutter (Chapter 3)

4. Draw attention where you want it (Chapter 4)

5. Think like a designer (Chapter 5)

6. Tell a story (Chapter 7)

In this chapter, we will look at the comprehensive storytelling with data process from start to end—applying each of the preceding lessons—using a single example.

Let's begin by considering Figure 8.1, which shows average retail price over time for five consumer products (A, B, C, D, and E). Spend a moment studying it.

FIGURE 8.1   Original visual

When presented with this graph, it's easy to start picking it apart. But before we discuss the best way to visualize the data shown in Figure 8.1, let's take a step back and consider the context.

Lesson 1: understand the context

The first thing to do when faced with a visualization challenge is to make sure you have a robust understanding of the context and what you need to communicate. We must identify a specific audience and what they need to know or do, and determine the data we'll use to illustrate our case. We should craft the Big Idea.

In this case, let's assume we work for a startup that has created a consumer product. We are starting to think about how to price the product. One of the considerations in this decision-making process—the one we will focus on here—is how competitors' retail prices for products in this marketplace have changed over time. There is an observation made with the original visual that may be important: "Price has declined for all products on the market since the launch of Product C in 2010."

If we pause to consider specifically the *who*, *what*, and *how*, let's assume following:

Who: VP of Product, the primary decision maker in establishing our product's price.

What: Understand how competitors' pricing has changed over time and recommend a price range.

How: Show average retail price over time for Products A, B, C, D, and E.

The Big Idea, then, could be something like: Based on analysis of pricing in the market over time, to be competitive, we recommend introducing our product at a retail price in the range $ABC–$XYZ.

Next, let's consider some different ways to visualize this data.

Lesson 2: choose an appropriate display

Once we've identified the data we want to show, next comes the challenge of determining how to best visualize it. In this case, we are most interested in the trend in price over time for each product. If we look back to Figure 8.1, the variance in colors across the bars distract from this, making the exercise more difficult than necessary. Bear with me, as we're going to go through more iterations of looking at this data than you might typically. The progression is interesting because it illustrates how different views of the data can influence what you pay attention to and the observations you can easily make.

First, let's remove the visual obstacle of the variance in color and view the resulting graph, shown in Figure 8.2.

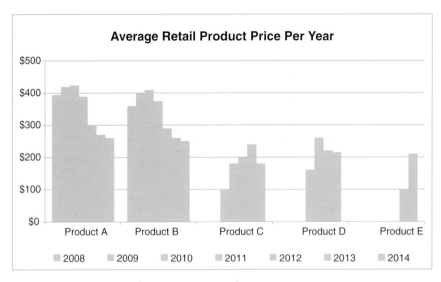

FIGURE 8.2 Remove the variance in color

If you're tempted to continue decluttering at this point, you are not alone. I had to resist the urge since that's something I typically do as I go along. In this case, let's refrain from doing so until the next section, where we can address it all at once.

Since the emphasis in the original headline was on what happened since Product C was launched in 2010, let's highlight the relevant pieces of data to make it easier to focus our attention there for a moment. See Figure 8.3.

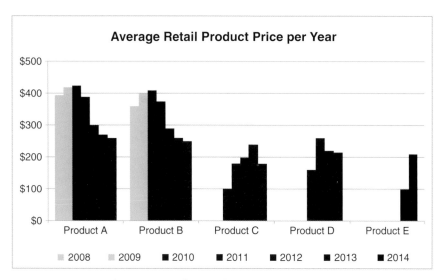

FIGURE 8.3 Emphasize 2010 forward

Upon studying this, we see clear declines in the average retail price for Products A and B in the time period of interest, but this doesn't appear to hold true for the products that were launched later. We will definitely need to change the headline from the original visual to reflect this when we tell our comprehensive story.

If you've been thinking we should try a line graph here instead of a bar chart—since we are primarily interested in the trend over time—you are absolutely right. In doing so, we also eliminate the stairstep view that bars create somewhat artificially. Let's see what lines would look like with the same layout as above. This is illustrated in Figure 8.4.

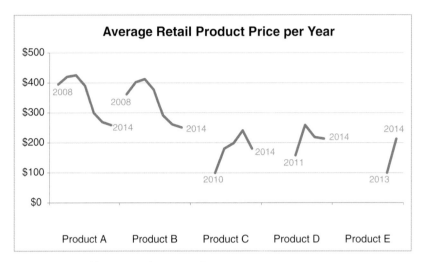

FIGURE 8.4   Change to line graph

The view in Figure 8.4 allows us to see what's happening over time more clearly for one product at a time. But it is hard to compare the products at a given point in time to one another. Graphing all of the lines against the same x-axis will solve this. This will also reduce the clutter and redundancy of the multiple year labels. The resulting graph might look like Figure 8.5.

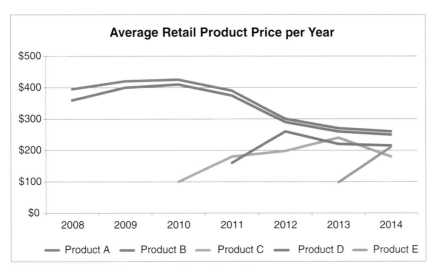

FIGURE 8.5   Single line graph for all products

With the transition to the new graph setup, Excel added back the color that we removed in an earlier step (tying the data to the accompanying legend at the bottom). Let's ignore that for a moment while we consider whether this view of the data will meet our needs. If we revisit our purpose, it is to understand how competitors' prices have changed over time. The way the data is shown in Figure 8.5 allows for this with relative ease. We can make taking in this information even easier by eliminating clutter and drawing attention where we want it.

Lesson 3: eliminate clutter

Figure 8.5 shows what our visual looks like when we rely on the default settings of our graphing application (Excel). We can improve this with the following changes:

- **De-emphasize the chart title.** It needs to be present, but doesn't need to attract as much attention as it does when written in bold black.

- **Remove chart border and gridlines,** which take up space without adding much value. Don't let unnecessary elements distract from your data!
- **Push the *x*- and *y*-axis lines and labels to the background** by making them grey. They shouldn't compete visually with the data. Modify the x-axis tick marks so they align with the data points.
- **Remove the variance in colors between the various lines.** We can use color more strategically, which we'll discuss further momentarily.
- **Label the lines directly,** eliminating the work of going back and forth between the legend and the data to understand what is being shown.

Figure 8.6 shows what the graph looks like after making these changes.

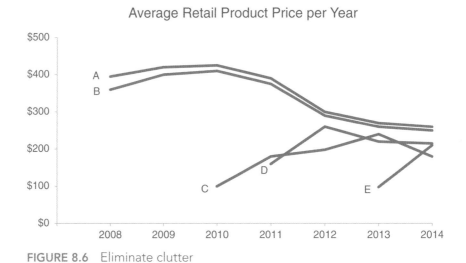

FIGURE 8.6 Eliminate clutter

Next, let's explore how we can focus our audience's attention.

Lesson 4: draw attention where you want your audience to focus

With the view shown in Figure 8.6, we can much more easily see and comment on what's happening over time. Let's explore how we can focus on different aspects of the data through strategic use of pre-attentive attributes.

Consider the initial headline: "Price has declined for all products on the market since the launch of Product C in 2010." Upon a closer look at the data, I might modify it to say something like, "After the launch of Product C in 2010, **the average retail price of existing products declined**." Figure 8.7 demonstrates how we can tie the important points in the data to these words through the strategic use of color.

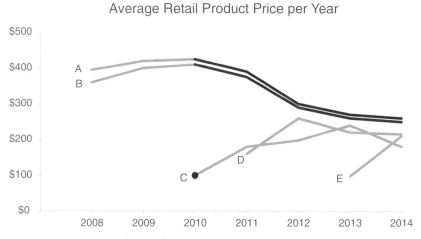

FIGURE 8.7   Focus the audience's attention

In addition to the colored segments of the lines in Figure 8.7, attention is also drawn to the introduction of Product C in 2010 through the addition of a data marker at that point. This is tied visually to the subsequent decrease over time in Products A and B through the consistent use of color.

Changing components of a graph in Excel

Typically, you format a series of data (a line or a series of bars) all at once. Sometimes, however, it can be useful to have certain points formatted differently—for example, to draw attention to specific parts, as illustrated in Figures 8.7, 8.8, and 8.9. To do this, click on the data series once to highlight it, then click again to highlight just the point of interest. Right-click and select Format Data Point to open the menu that will allow you to reformat the specific point as desired (for example, to change the color or add a data marker). Repeat this process for each data point you want to modify. It takes time, but the resulting visual is easier to comprehend for your audience. It is time well spent!

We can use this same view and strategy to concentrate on another observation—one perhaps more interesting and noteworthy: "With the launch of a new product in this space, it is typical to see an initial average retail price **increase**, followed by a **decline**." See Figure 8.8.

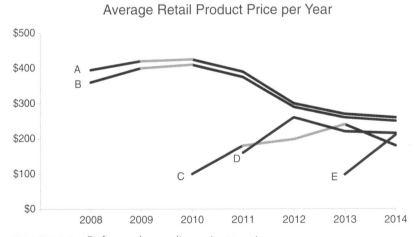

FIGURE 8.8   Refocus the audience's attention

It might also be interesting to note, "As of 2014, retail prices have converged across products, with an **average retail price of $223,** ranging from a low of $180 (Product C) to a high of $260 (Product A)." Figure 8.9 uses color and data markers to draw our attention to the specific points in the data that support this observation.

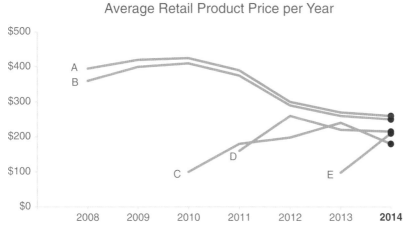

FIGURE 8.9   Refocus the audience's attention again

With each different view of the data, the use of preattentive attributes allows you to more clearly see certain things. This strategy can be used to highlight and tell different pieces of a nuanced story.

But before we continue thinking through how to best tell the story, let's put on our designer hats and perfect the visual.

Lesson 5: think like a designer

Though you may not have recognized it explicitly as such, we've already been thinking like a designer through this process. Form follows function: we chose a visual display (form) that will allow our audience to do what we need them to do (function) with ease. When it comes to using visual affordances to make it clear how our audience should interact with our visual, we've already taken steps to cut clutter and de-emphasize some elements of the graph, while emphasizing and drawing attention to others.

We can further improve this visual by leveraging the lessons we covered in Chapter 5 with respect to accessibility and aesthetics. Specifically, we can:

- **Make the visual accessible with text.** We can use simpler text in the graph title and capitalize only the first word to make it easier to comprehend and quicker to read. We also need to add axis titles to both the vertical and horizontal axes.
- **Align elements to improve aesthetics:** The center alignment of the graph title leaves it hanging in space and doesn't align it with any other elements; we should upper-left-most align the graph title. Align the y-axis title vertically with the uppermost label and the x-axis title horizontally with the leftmost label. This creates cleaner lines and ensures that your audience sees how to interpret what they are looking at before they get to the actual data.

Figure 8.10 shows what the visual looks like after these changes have been made.

Retail price over time

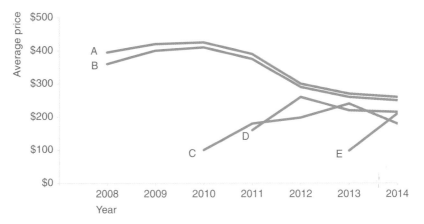

FIGURE 8.10   Add text and align elements

Lesson 6: tell a story

Finally, it is time to think about how we can use the visual we've created in Figure 8.10 as a foundation to walk our audience through the story in the way that we want them to experience it.

Imagine we have five minutes in a live presentation setting under the agenda topic: "Competitive Landscape—Pricing." The following sequence (Figures 8.11–8.19) illustrates one path we could take for telling a story with this data.

In the next **5 minutes**...

OUR GOAL:

1 Understand **how prices have changed over time** in the competitive landscape.

2 Use this knowledge to **inform the pricing of our product**.

We will end with a **specific recommendation**.

FIGURE 8.11

Products A and B were launched in 2008 at price points of **$360+**

Retail price over time

FIGURE 8.12

They have been priced similarly over time, with B consistently slightly lower than A

Retail price over time

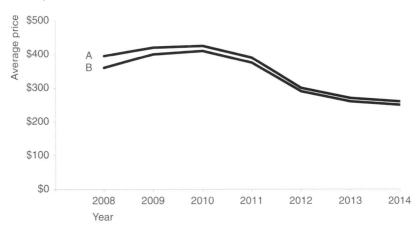

FIGURE 8.13

In 2014, Products A and B were priced at **$260** and **$250**, respectively

Retail price over time

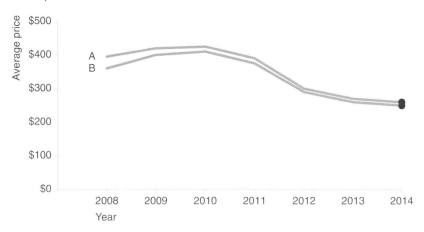

FIGURE 8.14

Products C, D, and E were each introduced later at **much lower price points**...

Retail price over time

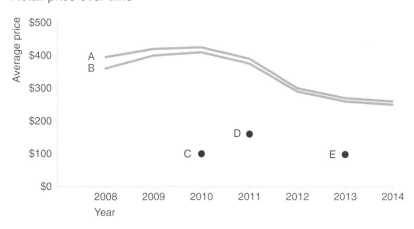

FIGURE 8.15

…but all have **increased in price** since their respective launches

Retail price over time

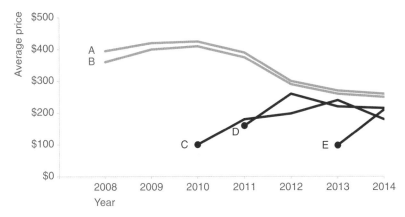

FIGURE 8.16

In fact, with the launch of a new product in this space, we tend to see an **initial price increase**, followed by a **decrease** over time

Retail price over time

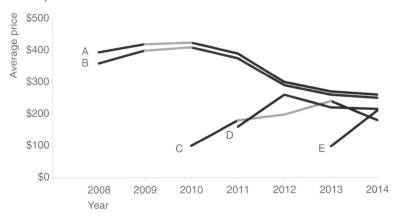

FIGURE 8.17

As of 2014, retail prices have converged, with an **average retail price of $223**, ranging from a low of $180 (C) to a high of $260 (A)

Retail price over time

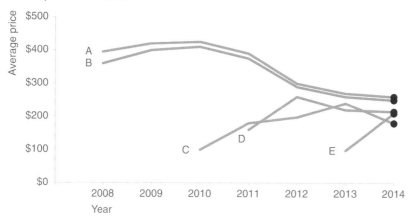

FIGURE 8.18

To be competitive, we recommend introducing our product *below the $223 average* price point in the **$150–$200 range**

Retail price over time

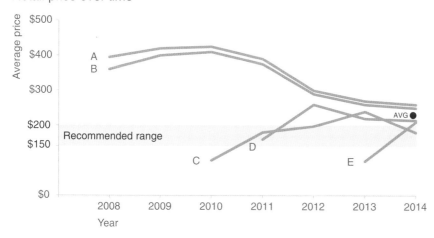

FIGURE 8.19

Let's consider this progression. We started off by telling our audience the structure we would follow. I can imagine the voiceover in the live presentation could further set the plot before moving to the next slide: "As you all know, there are five products that will be our key competition in the marketplace," then building the chronological price path that those products followed. We can introduce tension in the competitive landscape when Products C, D, and E significantly undercut existing price points at their respective launches. We can then restore a sense of balance as the prices converge. We end with a clear call to action: the recommendation for pricing our product.

By drawing our audience's attention to the specific part of the story we want to focus on—either by only showing the relevant points or by pushing other things to the background and emphasizing only the relevant pieces and pairing this with a thoughtful narrative—we've led our audience through the story.

Here, we've looked at an example telling a story with a single visual. This same process and individual lessons can be followed when you have multiple visuals in a broader presentation or communication. In that case, think about the overarching story that ties it all together. Individual stories for a given visualization within that larger presentation, such as the one we've looked at here, can be considered subplots within the broader storyline.

In closing

Through this example, we've seen the storytelling with data process from start to finish. We began by building a robust understanding of the context. We chose an appropriate visual display. We identified and eliminated clutter. We used preattentive attributes to draw our audience's attention to where we want them to focus. We put on our designer hats, adding text to make our visual accessible and employing alignment to improve the aesthetics. We crafted a compelling narrative and told a story.

Consider the before-and-after shown in Figure 8.20.

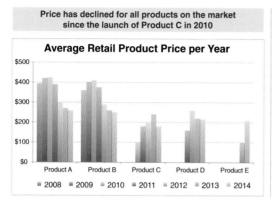

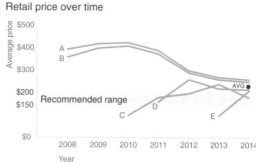

FIGURE 8.20 Before-and-after

The lessons we've learned and employed help us move from simply showing data to **storytelling with data**.

chapter nine

case studies

At this point, you should feel like you have a solid foundation for communicating effectively with data. In this penultimate chapter, we explore strategies for tackling common challenges faced when communicating with data through a number of case studies.

Specifically, we'll discuss:

- Color considerations with a dark background
- Leveraging animation in the visuals you present
- Establishing logic in order
- Strategies for avoiding the spaghetti graph
- Alternatives to pie charts

Within each of these case studies, I'll apply the various lessons we've covered when it comes to communicating effectively with data, but will limit my discussion mainly to the specific challenge at hand.

CASE STUDY 1: Color considerations with a dark background

When it comes to communicating data, I don't typically recommend anything other than a white background. Let's take a look at what a simple graph looks like on a white, blue, and black background. See Figure 9.1.

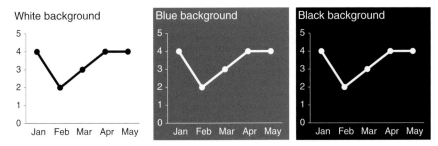

FIGURE 9.1   Simple graph on white, blue, and black background

If you had to describe in a single word how the blue and black backgrounds in Figure 9.1 make you feel, what would that word be? For me, it would be *heavy*. With the white background, I find it easy to focus on the data. The dark backgrounds, on the other hand, pull my eyes there—to the background—and away from the data. Light elements on a dark background can create a stronger contrast but are generally harder to read. Because of this, I typically avoid dark and colored backgrounds.

That said, sometimes there are considerations outside of the ideal scenario for communicating with data that must be taken into account, such as your company or client's brand and corresponding standard template. This was the challenge I faced in one consulting project.

I didn't recognize this immediately. It was only after I had completed my initial revamp of the client's original visual that I realized it just didn't quite fit with the look and feel of the work products I'd seen from the client group. Their template was bold and in your face with a mottled, black background spiked with bright, heavily saturated colors. In comparison, my visual felt rather meek. Figure 9.2 shows a generalized version of my initial makeover of a visual displaying employee survey feedback.

Survey Results: Team X

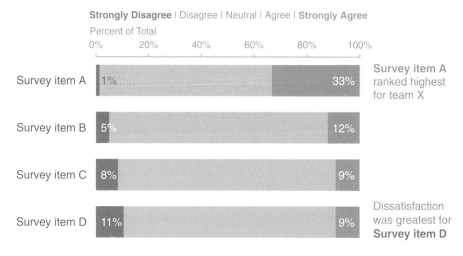

Strongly Disagree | Disagree | Neutral | Agree | **Strongly Agree**

Percent of Total

0%   20%   40%   60%   80%   100%

Survey item A   1%   33%   **Survey item A** ranked highest for team X

Survey item B   5%   12%

Survey item C   8%   9%

Survey item D   11%   9%   Dissatisfaction was greatest for **Survey item D**

FIGURE 9.2   Initial makeover on white background

In an endeavor to create something more in sync with the client's brand, I remade my own makeover, leveraging the same dark background I'd seen used in some of the other examples shared. In doing so, I had to reverse my normal thought process. With a white background, the further a color is from white, the more it will stand out (so grey stands out less, whereas black stands out very much). With a black background, the same is true, but black becomes the baseline (so grey stands out less, and white stands out very much). I also realized some colors that are typically verboten with a white background (for example, yellow) are incredibly attention grabbing against black (I didn't use yellow in this particular example but did in some others).

Figure 9.3 depicts how my "more in line with the client's brand" version of the visual looked.

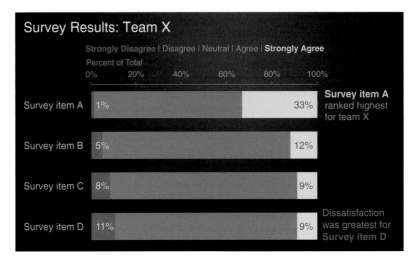

FIGURE 9.3   Remake on dark background

While the content is exactly the same, note how different Figure 9.3 feels compared to Figure 9.2. This is a good illustration of how color can impact the overall tone of a visualization.

CASE STUDY 2: Leveraging animation in the visuals you present

One conundrum commonly faced when communicating with data is when a single view of the data is used for both presentation and report. When presenting content in a live setting, you want to be able to walk your audience through the story, focusing on just the relevant part of the visual. However, the version that gets circulated to your audience—as a pre-read or takeaway, or for those who weren't able to attend the meeting—needs to be able to stand on its own without you, the presenter, there to walk the audience through it.

Too often, we use the exact same content and visuals for both purposes. This typically renders the content too detailed for the live presentation (particularly if it is being projected on the big screen) and sometimes not detailed enough for the circulated content. This gives rise to the slideument—part presentation, part document, and not exactly meeting the needs of either—which we touched upon briefly in Chapter 1. In the following, we'll look at a strategy for leveraging animation coupled with an annotated line graph to meet both the presentation and circulation needs.

Let's assume that you work for a company that makes online social games. You are interested in telling the story around how active users for a given game—let's call it Moonville—have grown over time.

You could use Figure 9.4 to talk about growth since the launch of the game in late 2013.

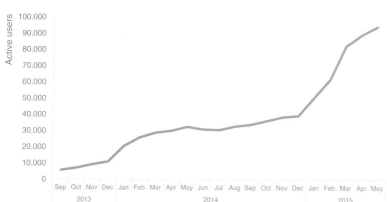

Moonville: active users over time

Data source: ABC Report. For purpose of analysis "active user" is defined as the number of unique users in the past 30 days.

FIGURE 9.4   Original graph

The challenge, however, is that when you put this much data in front of your audience, you lose control over their attention. You might be talking about one part of the data while they are focusing somewhere else entirely. Perhaps you want to tell the story chronologically, but your audience may jump immediately to the sharp increase in 2015 and wonder what drove that. When they do so, they stop listening to you.

Alternatively, you can leverage animation to walk your audience through your visual as you tell the corresponding points of the story. For example, I could start with a blank graph. This forces the audience to look at the graph details with you, rather than jump straight to the data and start trying to interpret it. You can use this approach to build anticipation within your audience that will help you to retain their attention. From there, I subsequently show or highlight *only the data that is relevant to the specific point I am making*, forcing the audience's attention to be exactly where I want it as I am speaking.

I might say—and show—the following progression:

Today, I'm going to talk you through a success story: the increase in Moonville users over time. First, let me set up what we are looking at. On the vertical y-axis of this graph, we're going to plot active users. This is defined as the number of unique users in the past 30 days. We'll look at how this has changed over time, from the launch in late 2013 to today, shown along the horizontal x-axis. (Figure 9.5)

Moonville: active users over time

Data source: ABC Report. For purpose of analysis "active user" is defined as the number of unique users in the past 30 days.

FIGURE 9.5

*We launched Moonville in September 2013. By the end of that first
month, we had just over 5,000 active users, denoted by the big blue
dot at the bottom left of the graph. (Figure 9.6)*

Moonville: active users over time

Data source: ABC Report. For purpose of analysis "active user" is defined as the number of unique users in the past 30 days.

FIGURE 9.6

Early feedback on the game was mixed. In spite of this—and our practically complete lack of marketing—the number of active users nearly doubled in the first four months, to almost 11,000 active users by the end of December. (Figure 9.7)

Moonville: active users over time

Data source: ABC Report. For purpose of analysis "active user" is defined as the number of unique users in the past 30 days.

FIGURE 9.7

In early 2014, the number of active users increased along a steeper trajectory. This was primarily the result of the friends and family promotions we ran during this time to increase awareness of the game. (Figure 9.8)

Moonville: active users over time

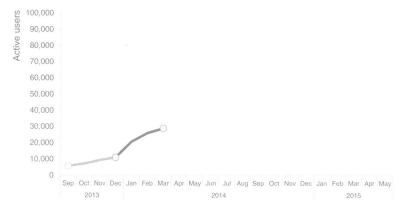

Data source: ABC Report. For purpose of analysis "active user" is defined as the number of unique users in the past 30 days.

FIGURE 9.8

Growth was pretty flat over the rest of 2014 as we halted all marketing efforts and focused on quality improvements to the game. (Figure 9.9)

Moonville: active users over time

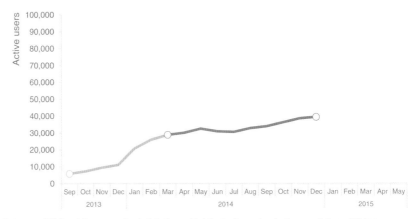

Data source: ABC Report. For purpose of analysis "active user" is defined as the number of unique users in the past 30 days.

FIGURE 9.9

Uptake this year, on the other hand, has been incredible, surpass-
ing our expectations. The revamped and improved game has gone
viral. The partnerships we've forged with social media channels have
proven successful for continuing to increase our active user base. At
recent growth rates, we anticipate we'll surpass 100,000 active users
in June! (Figure 9.10)

Moonville: active users over time

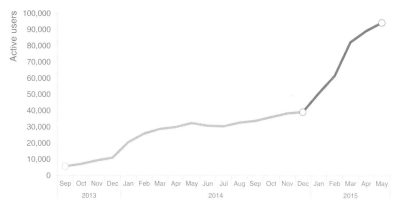

Data source: ABC Report. For purpose of analysis "active user" is defined as the number of unique users in the past 30 days.

FIGURE 9.10

For the more detailed version that you circulate as a follow up or for
those who missed your (stellar) presentation, you can leverage a version that annotates the salient points of the story on the line graph
directly, as shown in Figure 9.11.

Data source: ABC Report. For purpose of analysis "active user" is defined as the number of unique users in the past 30 days.

FIGURE 9.11

This is one strategy for creating a visual (or, in this case, set of visuals) that meets both the needs of your live presentation and the circulated version. Note that with this approach, it is imperative that you know your story well to be able to narrate without relying on your visuals (something you should always aim for regardless).

If you're leveraging presentation software, you can set up all of the above on a single slide and use animation for the live presentation, having each image appear and disappear as needed to form the desired progression. Put the final annotated version on top so it's all that shows on the printed version of the slide. If you do this, you can use the exact same deck for the presentation and the communication that you circulate. Alternatively, you can put each graph on a separate slide and flip through them; in this case, you'd only want to circulate the final annotated version.

CASE STUDY 3: Logic in order

There should be logic in the order in which you display information.

The above statement probably goes without saying. Yet, like so many things that seem logical when we read them or hear them or say them out loud, too often we don't put them into practice. This is one such example.

While I would say my introductory sentence is universally true, I'll focus here on a very specific example to illustrate the concept: leveraging order for categorical data in a horizontal bar chart.

First, let's set the context. Let's say you work at a company that sells a product that has various features. You've recently surveyed your users to understand whether they are using each of the features and how satisfied they've been with them and want to put that data to use. The initial graph you create might look something like Figure 9.12.

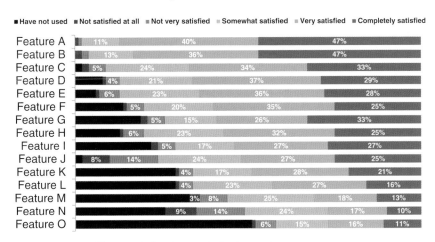

FIGURE 9.12   User satisfaction, original graph

This is a real example, and Figure 9.12 shows the actual graph that was created for this purpose, with the exception that I've replaced the descriptive feature names with Feature A, Feature B, and so on. There is an order here—if we stare at the data for a bit, we find that it is arranged in decreasing order of the "Very satisfied" group plus the "Completely satisfied" group (the teal and dark teal segments on the right side of the graph). This may suggest that is where we should pay attention. But from a color standpoint, my eyes are drawn first to the bold black "Have not used" segment. And if we pause to think about what the data shows, it would perhaps be the areas of dissatisfaction that would be of most interest.

Part of the challenge here is that the story—the "so what"—of this visual is missing. We could tell a number of different stories and focus on a number of different aspects of this data. Let's look at a couple of ways to do this, with an eye towards leveraging order.

First, we could think about highlighting the positive story: where our users are most satisfied. See Figure 9.13.

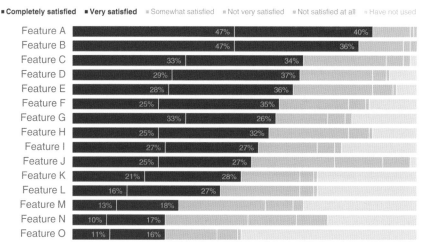

FIGURE 9.13   Highlight the positive story

In Figure 9.13, I've ordered the data clearly by putting "Completely satisfied" plus "Very satisfied" in descending order—the same as in the original graph—but I've made it much more obvious here through other visual cues (namely, color, but also the positioning of the segments as the first series in the graph, so the audience's attention hits it first as they scan from left to right). I've also used words to help explain *why* your attention is drawn to where it is via the action title at the top, which calls out what you should be seeing in the visual.

We can leverage these same tactics—order, color, placement, and words—to highlight a different story within this data: where users are least satisfied. See Figure 9.14.

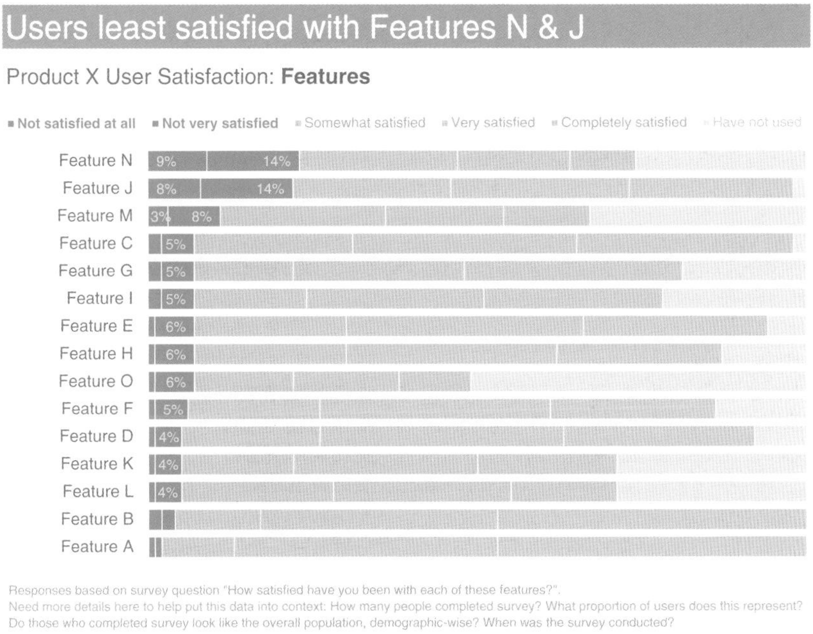

Users least satisfied with Features N & J

Product X User Satisfaction: **Features**

■ Not satisfied at all   ■ Not very satisfied   ■ Somewhat satisfied   ■ Very satisfied   ■ Completely satisfied   ■ Have not used

Feature N   9%   14%
Feature J   8%   14%
Feature M   3%   8%
Feature C   5%
Feature G   5%
Feature I   5%
Feature E   6%
Feature H   6%
Feature O   6%
Feature F   5%
Feature D   4%
Feature K   4%
Feature L   4%
Feature B
Feature A

Responses based on survey question "How satisfied have you been with each of these features?".
Need more details here to help put this data into context: How many people completed survey? What proportion of users does this represent?
Do those who completed survey look like the overall population, demographic-wise? When was the survey conducted?

FIGURE 9.14   Highlight dissatisfaction

Or perhaps the real story here is in the unused features, which could be highlighted as shown in Figure 9.15.

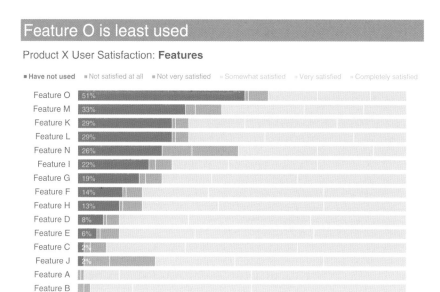

FIGURE 9.15   Focus on unused features

Note that in Figure 9.15, you can still get to the differing levels of satisfaction (or lack thereof) within each bar, but they've been pushed back to a second-order comparison due to the color choices I've made, while the relative rank ordering of the "Have not used" segment is the clear primary comparison on which my audience is meant to focus.

If we want to tell one of the above stories, we can leverage order, color, position, and words as I've shown to draw our audience's attention to where we want them to pay it in the data. If we want to tell *all three* stories, however, I'd recommend a slightly different approach.

It isn't very nice to get your audience familiar with the data only to completely rearrange it. Doing so creates a mental tax—the same sort of unnecessary cognitive burden that we discussed in Chapter 3 that we want to avoid. Let's create a base visual and preserve the same order so our audience only has to familiarize themselves with the detail once—highlighting the different stories one at a time through strategic use of color.

FIGURE 9.16   Set up the graph

Figure 9.16 depicts our base visual, without anything highlighted. If I were presenting this to an audience, I'd use this version to walk them through what they are looking at: survey responses to the question, "How satisfied have you been with each of these features?"—ranging from the positive "Completely satisfied" at the right to "Not satisfied at all" and, finally, "Have not used" at the far left (leveraging the natural association of positive at the right and negative at the left). Then I'd pause to tell each of the stories in succession.

First comes a visual similar to what we started with in the last series that highlights where users are the most satisfied. In this version, I've leveraged different shades of blue to draw attention not only to the proportion of users who are satisfied but specifically to Features A and B within those segments that rank highest, tying these bars visually to the text that illustrates my point. See Figure 9.17.

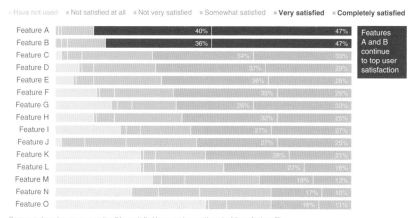

FIGURE 9.17   Satisfaction

This is followed by a focus on the other end of the spectrum to where users are least satisfied, again calling out and highlighting specific points of interest. See Figure 9.18.

FIGURE 9.18   Dissatisfaction

Note how it isn't as easy to see the relative rank ordering of the features highlighted in Figure 9.18 as it was when they were put in descending order (Figure 9.14) because they aren't aligned along a common baseline to either the left or the right. We can still relatively quickly see the primary areas of dissatisfaction (Features J and N) since they are so much bigger than the other categories and because of the color emphasis. I've added a callout box to highlight this through text as well.

Finally, preserving the same order, we can draw our audience's attention to the unused features. See Figure 9.19.

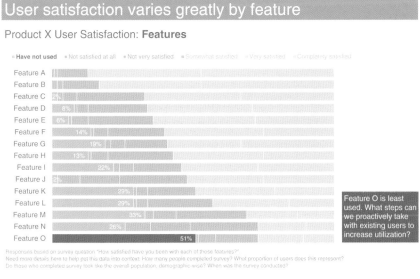

FIGURE 9.19   Unused features

In Figure 9.19, it is easier to see the rank ordering (even though the categories aren't monotonically increasing from top to bottom) because of the alignment to a consistent baseline at the left of the graph. Here, we want our audience to focus mainly on the very bottom feature in the graph—Feature O. Since we're trying to preserve the established order and can't do this by putting it at the top (where the audience would encounter it first), the bold color and callout box help draw attention to the bottom of the graph.

The preceding views show the progression I'd use in a live presentation. The sparing and strategic use of color lets me direct my audience's attention to one component of the data at a time. If you are creating a written document to be shared directly with your audience, you might compress all of these views into a single, comprehensive visual, as shown in Figure 9.20.

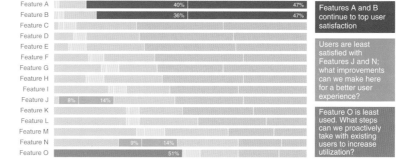

FIGURE 9.20   Comprehensive visual

When I process Figure 9.20, my eyes do a number of zigzagging "z's" across the page. First, I see the bold "Features" in the graph title. Then I'm drawn to the dark blue bars—which I follow across to the dark blue text box that tells me what's interesting about what I'm looking at (you'll note my text here is mostly descriptive, mainly due to the anonymity of the example; ideally this space would be used to provide greater insight). Next, I hit the orange text box, read it, and glance back leftward to see the evidence in the graph that supports it. Finally, I see the teal bar emphasized at the bottom and look across to see the text that describes it. Strategic use of color sets the various series apart from one another while also making it clear where the audience should look for the specific evidence of what is being described in the text.

Note that with Figure 9.20 it is harder for your audience to form *other* conclusions with the data, since attention is drawn so strongly to the particular points I want to highlight. But as we've discussed repeatedly, once you've reached the point of needing to communicate, *there should be a specific story or point that you want to highlight*, rather than let your audience draw their own conclusions. Figure 9.20 is too dense for a live presentation but could work well for the document that will be circulated.

I've mentioned this previously but would feel remiss not to point out that in some cases there is intrinsic order in the data you want to show (ordinal categories). For example, instead of features, if the categories were age ranges (0–9, 10–19, 20–29, etc.), you should keep those categories in numerical order. This provides an important construct for the audience to use as they interpret the information. Then use the other methods of drawing attention (through color, position, callout boxes with text) to direct the audience's attention to where you want them to pay it.

Bottom line: there should be logic in the order of the data you show.

CASE STUDY 4: Strategies for avoiding the spaghetti graph

While I very much enjoy food, I have a distaste for any chart type that has food in its title. My hatred of pie charts is well documented. Donuts are even worse. Here is another to add to the list: the spaghetti graph.

If you aren't sure if you've seen a spaghetti graph before, I'll bet that you have. A spaghetti graph is a line graph where the lines overlap a lot, making it difficult to focus on a single series at a time. They look something like Figure 9.21.

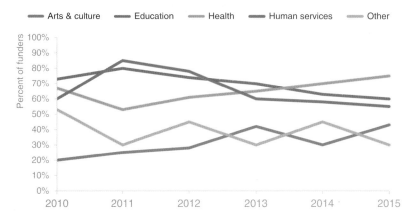

Types of non-profits supported by area funders

Data is self-reported by funders; percents sum to greater than 100 because respondents can make multiple selections.

FIGURE 9.21   The spaghetti graph

Graphs like Figure 9.21 are known as spaghetti graphs because they look like someone took a handful of uncooked spaghetti noodles and threw them on the ground. And they are about as informative as those haphazard noodles would be as well …

which is to say …

not at all.

Note how difficult it is to concentrate on a single line within that mess, due to all of the crisscrossing and because so much is competing for your attention.

There are a few strategies for taking the would-be-spaghetti graph and creating more visual sense of the data. I'll cover three such strategies and show them applied in a couple of different ways to the data graphed in Figure 9.21, which shows types of nonprofits supported by funders in a given area. First, we'll look at an approach you should be familiar with by now: using preattentive attributes to emphasize a single line at a time. After that, we'll look at a couple of views that separate the lines spatially. Then finally, we'll look at a combined approach that leverages elements of these first two strategies.

Emphasize one line at a time

One way to keep the spaghetti graph from becoming visually overwhelming is to use preattentive attributes to draw attention to a single line at a time. For example, we could focus our audience on the increase in the percentage of funders donating over time to health nonprofits. See Figure 9.22.

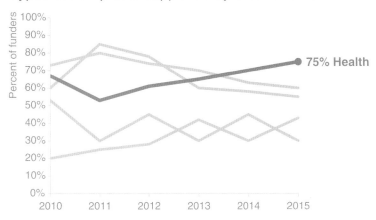

Types of non-profits supported by area funders

Data is self-reported by funders; percents sum to greater than 100 because respondents can make multiple selections.

FIGURE 9.22   Emphasize a single line

Or we could use the same strategy to emphasize the decrease in the percentage of funders donating to education-related nonprofits. See Figure 9.23.

Types of non-profits supported by area funders

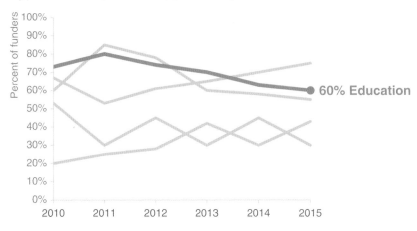

Data is self-reported by funders; percents sum to greater than 100 because respondents can make multiple selections.

FIGURE 9.23   Emphasize another single line

In Figures 9.22 and 9.23, color, thickness of line, and added marks (the data marker and data label) act as visual cues to draw attention to where we want our audience to focus. This strategy can work well in a live presentation, where you explain the details of the graph once (as we've seen in the recent case studies), then cycle through the various data series in this manner, highlighting what is interesting or should be paid attention to with each and why. Note that we need either this voiceover or the addition of text to make it clear why we are highlighting the given data and provide the story for our audience.

Separate spatially

We can untangle the spaghetti graph by pulling the lines apart either vertically or horizontally. First, let's look at a version where the lines are pulled apart vertically. See Figure 9.24.

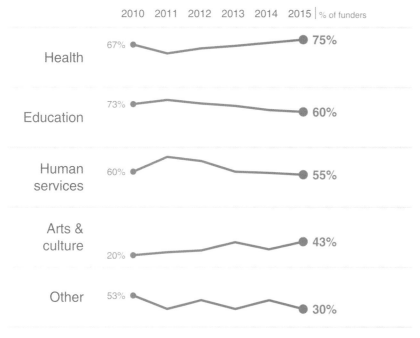

Types of non-profits supported by area funders

Data is self-reported by funders: percents sum to greater than 100 because respondents can make multiple selections.

FIGURE 9.24 Pull the lines apart vertically

In Figure 9.24, the same x-axis (year, shown at the top) is leveraged across all of the graphs. In this solution, I've created five separate graphs but organized them such that they appear to be a single visual. The y-axis within each graph isn't shown; rather, the starting and ending point labels are meant to provide enough context so that the axis is unnecessary. Though they aren't shown, it is important that the y-axis minimum and maximum are the same for each graph so the audience can compare the relative position of each line or point within the given space. If you were to shrink these down, they would look similar to what Edward Tufte calls "sparklines" (a very small line graph typically drawn without axis or coordinates to show the general shape of the data; *Beautiful Evidence*, 2006).

This approach assumes that being able to see the trend for a given category (Health, Education, etc.) is more important than comparing the values across categories. If that isn't the case, we can consider pulling the data apart horizontally, as illustrated in Figure 9.25.

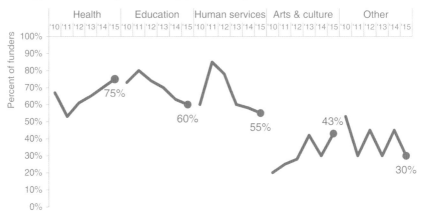

FIGURE 9.25   Pull the lines apart horizontally

Whereas in Figure 9.24 we leveraged the x-axis (years) across the five categories, in Figure 9.25 we leverage the same y-axis (percent of funders) across the five categories. Here, the relative height of the various data series allows them to more easily be compared with each other. We can quickly see that the highest percentage of funders in 2015 donate to Health, a lower percentage to Education, an even lower percentage to Human Services, and so on.

Combined approach

Another option is to combine the approaches we've outlined so far. We can separate spatially *and* emphasize a single line at a time, while leaving the others there for comparison but pushing them to the background. As was the case with the prior approach, we can do this by separating the lines vertically (Figure 9.26) or horizontally (Figure 9.27).

Types of non-profits supported by area funders

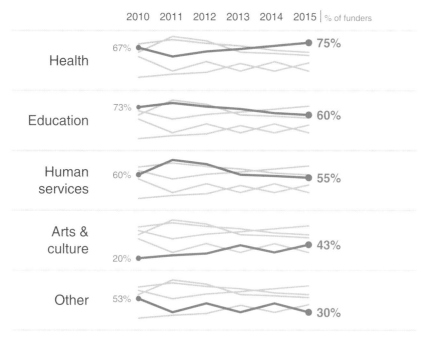

Data is self-reported by funders; percents sum to greater than 100 because respondents can make multiple selections.

FIGURE 9.26   Combined approach, with vertical separation

Types of non-profits supported by area funders

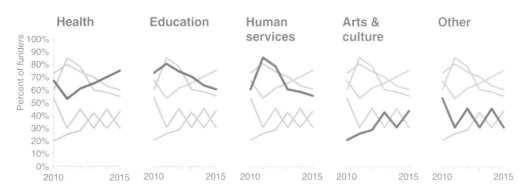

Data is self-reported by funders; percents sum to greater than 100 because respondents can make multiple selections.

FIGURE 9.27   Combined approach, with horizontal separation

Having a number of small graphs together, as shown in Figure 9.27, is sometimes referred to as "small multiples." As noted previously, it's imperative here that the details of each graph (the x- and y-axis minimum and maximum) are the same so that the audience can quickly compare the highlighted series across the various graphs.

This approach, shown in Figures 9.26 and 9.27, can work well if the context of the full dataset is important but you want to be able to focus on a single line at a time. Because of the denseness of information, this combined approach may work better for a report or presentation that will be circulated rather than a live presentation, where it will be more challenging to direct your audience where you want them to look.

As is frequently the case, there is not a single "right" answer. Rather, the solution that will work best will vary by situation. The meta-lesson is: if you find yourself facing a spaghetti graph, don't stop there. Think about what information you want to most convey, what story you want to tell, and what changes to the visual could help you accomplish that effectively. Note that in some cases, this may mean showing less data altogether. Ask yourself: Do I need all categories? All years? When appropriate, reducing the amount of data shown can make the challenge of graphing data like that shown in this example easier as well.

CASE STUDY 5: Alternatives to pies

Recall the scenario we discussed in Chapter 1 about the summer learning program on science. To refresh your memory: you just completed a pilot summer program on science aimed at improving perceptions of the field among 2nd and 3rd grade elementary children. You conducted a survey going into the program and at the end of the program, and want to use this data as evidence of the success of the pilot program in your request for future funding. Figure 9.28 shows a first attempt at graphing this data.

Survey results: summer learning program on science

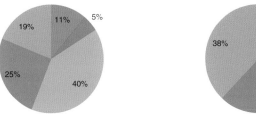

FIGURE 9.28   Original visual

The survey data demonstrates that, on the basis of improved sentiment toward science, the pilot program was a great success. Going into the program, the biggest segment of students (40%, the green slice in Figure 9.28, left) felt just "OK" about science—perhaps they hadn't made up their minds one way or the other. However, after the program (Figure 9.28, right), we see the 40% in green shrinks down to 14%. "Bored" (blue) and "Not great" (red) went up a percentage point each, but the majority of the change was in a positive direction. After the program, nearly 70% of kids (purple plus teal segments) expressed some level of interest toward science.

Figure 9.28 does this story a great disservice. I shared my less-than-favorable view on pie charts in Chapter 2, so I hope this judgment is not met with surprise. Yes, you can get to the story from Figure 9.28, but you have to work for it and overcome the annoyance of trying to compare segments across two pies. As we've discussed, we want to limit or eliminate the work your audience has to do to get at the information, and we certainly don't want to annoy them. We can avoid such challenges by choosing a different type of visual.

Let's take a look at four alternatives for displaying this data—show the numbers directly, simple bar graph, stacked horizontal bar graph, and slopegraph—and discuss some considerations with each.

Alternative #1: show the numbers directly

If the improvement in positive sentiment is the main message we want to impart to our audience, we can consider making that the only thing we communicate. See Figure 9.29.

Pilot program was a success

After the pilot program,

68%

of kids expressed interest towards science,
compared to 44% going into the program.

Based on survey of 100 students conducted before and after pilot program (100% response rate on both surveys).

FIGURE 9.29   Show the numbers directly

Too often, we think we have to include all of the data and overlook the simplicity and power of communicating with just one or two numbers directly, as demonstrated in Figure 9.29. That said, if you feel you need to show more, look to one of the following alternatives.

Alternative #2: simple bar graph

When you want to compare two things, you should generally put those two things as close together as possible and align them along a common baseline to make this comparison easy. The simple bar graph does this by aligning the Before and After survey responses with a consistent baseline at the bottom of the graph. See Figure 9.30.

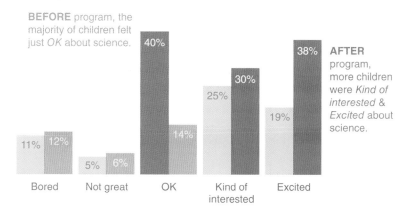

Pilot program was a success

How do you feel about science?

BEFORE program, the majority of children felt just *OK* about science.

40%

38% AFTER program, more children were *Kind of interested* & *Excited* about science.

30%

25%

19%

11% 12%

14%

5% 6%

| Bored | Not great | OK | Kind of interested | Excited |

Based on survey of 100 students conducted before and after pilot program (100% response rate on both surveys).

FIGURE 9.30   Simple bar graph

I am partial to this view for this specific example because the layout makes it possible to put the text boxes right next to the data points they describe (note that other data is there for context but is slightly pushed to the background through the use of lighter colors). Also, by having Before and After as the primary classification, I'm able to limit the visual to two colors—grey and blue—whereas three colors will be used in the following alternatives.

Alternative #3: 100% stacked horizontal bar graph

When the part-to-whole concept is important (something you don't get with either Alternative #1 or #2), the stacked 100% horizontal bar graph achieves this. See Figure 9.31. Here, you get a consistent baseline to use for comparison at the left and at the right of the graph. This allows the audience to easily compare both the negative segments at the left and the positive segments at the right across the two bars and, because of this, is a useful way to visualize survey data in general.

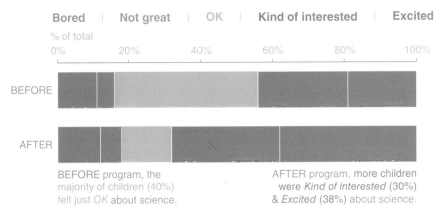

Pilot program was a success

How do you feel about science?

Bored | Not great | OK | Kind of interested | Excited

% of total

0%   20%   40%   60%   80%   100%

BEFORE

AFTER

BEFORE program, the majority of children (40%) felt just OK about science.

AFTER program, more children were *Kind of interested* (30%) & *Excited* (38%) about science.

Based on survey of 100 students conducted before and after pilot program (100% response rate on both surveys).

FIGURE 9.31   100% stacked horizontal bar graph

In Figure 9.31, I chose to retain the x-axis labels rather than put data labels on the bars directly. I tend to do it this way when leveraging 100% stacked bars so that you can use the scale at the top to read either from left to right or from right to left. In this case, it allows us to attribute numbers to the change from Before to After on the negative end of the scale ("Bored" and "Not great") or from right to left, doing the same for the positive end of the scale ("Kind of interested" and "Excited"). In the simple bar graph shown previously (Figure 9.30), I chose to omit the axis and label the bars directly. This illustrates how different views of your data may lead you to different design choices. Always think about how you want your audience to use the graph and make your design choices accordingly: different choices will make sense in different situations.

Alternative #4: slopegraph

The final alternative I'll present here is a slopegraph. As was the case with the simple bar chart, you don't get a clear sense of there being a whole and thus pieces-of-a-whole in this view (in the way that you do with the initial pie or with the 100% horizontal stacked bar). Also, if it is important to have your categories ordered in a certain way, a slopegraph won't always be ideal since the various categories are placed according to the respective data values. In Figure 9.32 on the right-hand side, you do get the positive end of the scale at the top, but note that "Bored" and "Not great" at the bottom are switched relative to how they'd appear in an ordinal scale because of the values that correspond with these points. If you need to dictate the category order, use the simple bar graph or the 100% stacked bar graph, where you can control this.

Pilot program was a success

How do you feel about science?

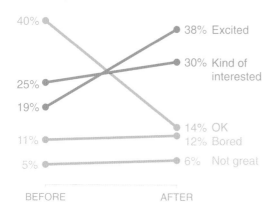

BEFORE program, the majority of children felt just *OK* about science.

AFTER program, more children were *Kind of interested* & *Excited* about science.

Based on survey of 100 students conducted before and after pilot program (100% response rate on both surveys).

FIGURE 9.32   Slopegraph

With the slopegraph in Figure 9.32, you can easily see the visual percentage change from Before to After for each category via the slope of the respective line. It's easy to see quickly that the category that increased the most was "Excited" (due to the steep slope) and the category that decreased markedly was "OK." The slopegraph also provides clear visual ordering of categories from greatest to least (via their respective points in space from top to bottom on the left and right sides of the graph).

Any of these alternatives might be the best choice given the specific situation, how you want your audience to interact with the information, and what point or points of emphasis you want to make. The big lesson here is that you have a number of alternatives to pies that can be more effective for getting your point across.

In closing

In this chapter, we discussed considerations and solutions for tackling several common challenges faced when communicating visually with data. Inevitably, you'll face data visualization challenges that I have not addressed. There is as much to be learned from the critical thinking that goes into solving some of these scenarios as there is from the "answer" itself. As we've discussed, when it comes to data visualization, rarely is there a single correct path or solution.

Even more examples

For more case studies like the ones we've considered here, check out my blog at storytellingwithdata.com, where you'll find a number of before-and-after examples leveraging the lessons that we've learned.

When you find yourself in a situation where you are unsure how to proceed, I nearly always recommend the same strategy: pause to consider your audience. What do you need them to know or do? What story do you aim to tell them? Often, by answering these questions, a good path for how to present your data will become clear. If one doesn't, try several views and seek feedback.

My challenge to you is to consider how you can apply all of the lessons we've learned and your critical thinking skills to the various and varied data visualization challenges you face. The responsibility—and the opportunity—to tell a story with data is yours.

chapter ten

final thoughts

Data visualization—and communicating with data in general—sits at the intersection of science and art. There is certainly some science to it: best practices and guidelines to follow, as we've discussed throughout this book. But there is also an artistic component. This is one of the reasons this area is so much fun. It is inherently diverse. Different people will approach things in varying ways and come up with distinct solutions to the same data visualization challenge. As we've discussed, there is no single "right" answer. Rather, there are often multiple potential paths for communicating effectively with data. Apply the lessons we've covered in this book to forge *your* path, with the goal of using your artistic license to make the information easier for your audience to understand.

You have learned a great deal over the course of this book that sets you up for success when it comes to communicating effectively with data. In this final chapter, we'll discuss some tips on where to go from here and strategies for upskilling storytelling with data competency in your team and organization. Finally, we will end with a recap of the main lessons we've covered and send you off eager and ready to tell stories with data.

Where to go from here

Reading about effective storytelling with data is one thing. But how do you translate what we've learned to practical application? The simple way to get good at this is to *do it*: practice, practice, and practice some more. Look for opportunities in your work to apply the lessons we've learned. Note that it doesn't have to be all or nothing—one way to make progress is through incremental improvements to existing or ongoing work. Consider also when you can leverage the entire storytelling with data process that we've covered from start to finish.

Now I want to overhaul our entire monthly report!

You likely see graphs differently than you did at the onset of our journey together. Rethinking the way you visualize data is a great thing. But don't let overambitious goals overwhelm and hinder progress. Consider what incremental improvements you can make as you work toward storytelling with data nirvana. For example, if you're considering overhauling your regular reports, an interim step could be to start thinking of the report as the appendix. Leave the data there for reference, but push it to the back so it doesn't distract from the main message. Insert a few slides or a cover note up front and use this to pull out the interesting stories, leveraging the storytelling with data lessons we've covered. This way you can more easily focus your audience on the important stories and resulting actions.

For some specific, concrete steps on where to go from here, I'll outline five final tips: learn your tools well, iterate and seek feedback, allow ample time for this part of the process, seek inspiration from others, and—last but not least—have some fun while you're at it! Let's discuss each of these.

Tip #1: learn your tools well

For the most part, I've intentionally avoided discussion on tools because the lessons we've covered are fundamental and can be applied to varying degrees in any tool (for example, Excel or Tableau). Try not to let your tools be a limiting factor when it comes to communicating effectively with data. Pick one and get to know it as best you can. When you're first starting out, a course to become familiar with the basics may be helpful. In my experience, however, the best way to learn a tool is to use it. When you can't figure out how to do something, don't give up. Continue to play with the program and search Google for solutions. Any frustration you encounter will be worth it when you can bend your tool to your will!

You don't need fancy tools in order to visualize data well. The examples we've looked at in this book were all created with Microsoft Excel, which I find is the most pervasive when it comes to business analytics.

While I use mainly Excel for visualizing data, this isn't your only option. There are a plethora of tools out there. The following is a very quick rundown of some of the popular ones currently used for creating data visualizations like the ones we've examined:

- **Google spreadsheets** are free, online, and sharable, allowing multiple people to edit (as of this writing, there remain graph formatting constraints that make it challenging to apply some of the lessons we've covered when it comes to decluttering and drawing attention where you want it).
- **Tableau** is a popular out-of-the-box data visualization solution that can be great for exploratory analysis because it allows you to quickly create multiple views and nice-looking graphs from your data. It can be leveraged for the explanatory via the Story Points

feature. It is expensive, though a free Tableau Public option is available if uploading your data to a public server isn't an issue.

- Programming languages—like **R, D3** (JavaScript), **Processing**, and **Python**—have a steeper learning curve but allow for greater flexibility, since you can control the specific elements of the graphs you create and make those specifications repeatable through code.
- Some people use **Adobe Illustrator**, either alone or together with graphs created in an application like Excel or via a programming language, for easier manipulation of graph elements and a professional look and feel.

How I use PowerPoint

For me, PowerPoint is simply the mechanism that allows me to organize a handout or present on the big screen. I nearly always start from a totally blank slide and do not leverage the built-in bullets that too easily turn content from presentation to teleprompter.

You can build graphs directly in PowerPoint; however, I tend not to do this. There is greater flexibility in Excel (where, in addition to the graph, you can also have some elements of a visual—for example, titles or axis labels—directly in the cells, which is sometimes useful). Because of this, I create my visuals in Excel, then copy and paste into PowerPoint as an image. If I am using text together with a visual—for example, to draw attention to a specific point—I typically do that via a text box in PowerPoint.

The animation feature within PowerPoint can be useful for progressing through a story with iterations of the same visual, as shown in Chapter 8 or some of the case studies in Chapter 9. When using animation in PowerPoint, use only simple Appear or Disappear (in some instances, Transparency can also be useful); steer clear of any animation that causes elements to fly in or fade out—this is the presentation software equivalent of 3D graphs—unnecessary and distracting!

Another essential basic tool for visualizing data that I did not include in the preceding list is **paper**—which brings me to my next tip.

Tip #2: iterate and seek feedback

I've presented the storytelling with data process as a linear path. That's not often the case in reality. Rather, it takes iterating to get from early ideas to a final solution. When the best course for visualizing certain data is unclear, start with a blank piece of paper. This enables you to brainstorm without the constraints of your tools or what you know how to do in your tools. Sketch out potential views to see them side-by-side and determine what will work best for getting your message across to your audience. I find that we form less attachment to our work product—which can make iterating easier—when we are working on paper rather than on our computers. There is also something freeing about drawing on blank paper that may make it easier to identify new approaches if you're feeling stuck. Once you have your basic approach sketched, consider what you have at your disposal—tools, or internal or external experts—to actually create the visual.

When creating your visual in your graphing application (for example, Excel) and refining to get from good to great, you can leverage what I call the "optometrist approach." Create a version of the graph (let's call it A), then make a copy of it (B) and make a single change. Then determine which looks better—A or B. Often, the practice of seeing slight variations next to each other makes it quickly clear which view is superior. Progress in this manner, preserving the latest "best" visual and continuing to make minor modifications in a copy (so you always have the prior version to go back to in case the modification worsens it) to iterate toward your ideal visual.

At any point, if the best path is unclear, seek feedback. The fresh set of eyes that a friend or colleague can bring to the data visualization endeavor is invaluable. Show someone else your visual and have them talk you through their thought process: what they pay attention to, what observations they make, what questions they have, and any ideas they may have for better getting your point across. These insights will let you know if the visual you've created is on the mark or, in the case when it isn't, give you an idea of where to make changes and focus continued iteration.

When it comes to iterating, there is one thing you need perhaps more than anything else in order to be successful: *time.*

Tip #3: devote time to storytelling with data

Everything we've discussed throughout this book takes time. It takes time to build a robust understanding of the context, time to under-stand what motivates our audience, time to craft the 3-minute story and form the Big Idea. It takes time to look at the data in different ways and determine how to best show it. It takes time to declutter and draw attention and iterate and seek feedback and iterate some more to create an effective visual. It takes time to pull it all together into a story and form a cohesive and captivating narrative.

It takes even more time to do all of this well.

One of my biggest tips for success in storytelling with data is to allow adequate time for it. If we don't consciously recognize that this takes time to do well and budget accordingly, our time can be entirely eaten up by the other parts of the analytical process. Consider the typical analytical process: you start with a question or hypothesis, then you collect the data, then you clean the data, and then you ana-lyze the data. After all of that, it can be tempting to simply throw the data into a graph and call it "done."

But we simply aren't doing ourselves—or our data—justice with this approach. The default settings of our graphing application are typi-cally far from ideal. Our tools do not know the story we aim to tell. Combine these two things and you run the risk of losing a great deal of potential value—including the opportunity to drive action and effect change—if adequate time isn't spent on this final step in the analytical process: the communication step. This is the only part of the entire process that your audience actually *sees*. Devote time to this important step. Expect it to take longer than you think to allow sufficient time to iterate and get it right.

Tip #4: seek inspiration through good examples

Imitation really is the best form of flattery. If you see a data visualization or example of storytelling with data that you like, consider how you might adapt the approach for your own use. Pause to reflect on what makes it effective. Make a copy of it and create a visual library that you can add to over time and refer to for inspiration. Emulate the good examples and approaches that you see.

Said more provocatively—imitation *is a good thing.* We learn by emulating experts. That's why you see people with their sketchpads and easels at art museums—they are interpreting great works. My husband tells me that while learning to play the jazz saxophone, he would listen to the masters repeatedly—narrowing at times to a single measure played at a slower speed that he would practice until he could repeat the notes perfectly. This idea of using great examples as an archetype to learn applies to data visualization as well.

There are a number of great blogs and resources on the topic of data visualization and communicating with data that contain many good examples. Here are a few of my current personal favorites (including my own!):

- **Eager Eyes** (eagereyes.org, Robert Kosara): Thoughtful content on data visualization and visual storytelling.
- **FiveThirtyEight's Data Lab** (fivethirtyeight.com/datalab, various authors): I like their typically minimalist graphing style on a large range of news and current events topics.
- **Flowing Data** (flowingdata.com, Nathan Yau): Membership gets you premium content, but there are a lot of great free examples of data visualization as well.

- **The Functional Art** (thefunctionalart.com, Alberto Cairo): An introduction to information graphics and visualization, with great concise posts highlighting advice and examples.
- **The Guardian Data Blog** (theguardian.com/data, various authors): News-related data, often with accompanying article and visualizations, by the British news outlet.
- **HelpMeViz** (HelpMeViz.com, Jon Schwabish): "Helping people with everyday visualizations," this site allows you to submit a visual to receive feedback from readers or scan the archives for examples and corresponding conversations.
- **Junk Charts** (junkcharts.typepad.com, Kaiser Fung): By self-proclaimed "web's first data viz critic," focuses on what makes graphics work and how to make them better.
- **Make a Powerful Point** (makeapowerfulpoint.com, Gavin McMahon): Fun, easy-to-digest content on creating and giving presentations and presenting data.
- **Perceptual Edge** (perceptualedge.com, Stephen Few): No-nonsense content on data visualization for sensemaking and communication.
- **Visualising Data** (visualisingdata.com, Andy Kirk): Charts the development of the data visualization field, with great monthly "best visualisations of the web" resource list.
- **VizWiz** (vizwiz.blogspot.com, Andy Kriebel): Data visualization best practices, methods for improving existing work, and tips and tricks for using Tableau Software.
- **storytelling with data** (storytellingwithdata.com): My blog focuses on communicating effectively with data and contains many examples, visual makeovers, and ongoing dialogue.

This is just a sampling. There is a lot of great content out there. I continue to learn from others who are active in this space and doing great work. You can, too!

Learn from the not-so-great examples, too

Often, you can learn as much from the poor examples of data visualization—what not to do—as you can from those that are effective. Bad graphs are so plentiful that entire sites exist to curate, critique, and poke fun at them. For an entertaining example, check out WTF Visualizations (wtfviz.net), where content is described simply as "visualizations that make no sense." I challenge you not only to recognize when you encounter a poor example of data visualization but also to pause and reflect on why it isn't ideal and how it could be improved.

You now have a discerning eye when it comes to the visual display of information. You will never look at a graph the same. One workshop attendee told me that he is "ruined"—he can't encounter a data visualization without applying his new lens for assessing effectiveness. I love hearing these stories, as it means I'm making progress toward my goal of ridding the world of ineffective graphs. You have been ruined in this same way, but this is actually a really good thing! Continue to learn from and leverage the aspects of good examples you see, while avoiding the pitfalls of the poor ones, as you start to create your own data visualization style.

Tip #5: have fun and find your style

When most people think about data, one of the furthest things from their mind is creativity. But within data visualization, there is absolutely space for creativity to play a role. Data can be made to be breathtakingly beautiful. Don't be afraid to try new approaches and play a little. You'll continue to learn what works and what doesn't over time.

You may also find that you develop a personal data visualization style. For example, my husband says he can recognize visuals that I created or influenced. Unless a client brand calls for something else, I tend to do everything in shades of grey and use blue sparingly in a minimalist style, almost always in plain old Arial font (I like it!). That doesn't mean your approach must imitate these specifics to be successful. My own style has evolved based on personal preferences and learning through trial and error—testing out different fonts, colors, and graph elements. I can recall one particularly unfortunate example that incorporated a grey-to-white shaded graph background and far too many shades of orange. I've come a long way!

To the extent that it makes sense given the task at hand, don't be afraid to let your own style develop and creativity come through when you communicate with data. Company brand can also play a role in developing a data visualization style; consider your company's brand and whether there are opportunities to fold that into how you visualize and communicate with data. Just make sure that your approach and stylistic elements are making the information easier—not more difficult—for your audience to consume.

Now that we've looked at some specific tips for *you* to follow, let's turn to some ideas for building storytelling with data competency in others.

Building storytelling with data competency in your team or organization

I am a strong believer that anyone can improve their ability to communicate with data by learning and applying the lessons we've covered. That said, some will have more interest and natural aptitude than others in this space. When it comes to being effective at communicating with data in your team or your organization, there are a few potential strategies to consider: upskill everyone, invest in an expert, or outsource this part of the process. Let's briefly discuss each of these.

Upskill everyone

As we've discussed, part of the challenge is that data visualization is a single step in the analytical process. Those hired into analytical roles typically have quantitative backgrounds that suit them well for the other steps (finding the data, pulling it together, analyzing it, building models), but not necessarily any formal training in design to help them when it comes to the communication of the analysis. Also, increasingly those without analytical backgrounds are being asked to put on analytical hats and communicate using data.

For both of these groups, finding ways to impart foundational knowledge can make everyone better. Invest in training or use the lessons covered here to generate momentum. On this latter note, here are some specific ideas:

- **Storytelling with data book club:** read a chapter at a time and then discuss it together, identifying examples specific to your work where the given lesson can be applied.
- **Do-it-yourself workshop:** after finishing the book, conduct your own workshop—soliciting examples of communicating with data from your team and discussing how they can be improved.
- **Makeover Monday:** challenge individuals to a weekly makeover of less-than-ideal examples employing the lessons we've covered.
- **Feedback loop:** set the expectation that individuals must share work in progress and offer feedback to each other grounded in the storytelling with data lessons.
- **And the winner is:** introduce a monthly or quarterly contest, where individuals or teams can submit their own examples of effective storytelling with data then start a gallery of model examples, adding to it over time via contest winners.

Any of these approaches—alone or combined—can create and help ensure continued focus on effective visualization and storytelling with data.

Invest in an internal expert or two

Another approach is to identify an individual or a couple of individuals on your team or in your organization who are interested in data visualization (even better if they've already displayed some natural aptitude) and invest in them so they can become your in-house experts. Make it an expectation of their role to be an internal data visualization consultant to whom others on the team can turn for brainstorming and feedback or to overcome tool-specific challenges. This investment can take the form of books, tools, coaching, workshops, or courses. Provide time and opportunities to learn and practice. This can be a great form of recognition and career development for the individual. As the individual continues to learn, they can share this with others as a way to ensure continued team development as well.

Outsource

In some situations, it may make sense to outsource visual creation to an external expert. If time or skill constraints are too great to overcome for a specific need, turning to a data visualization or presentation consultant may be worth considering. For example, one client contracted me to design an important presentation that they would need to give a number of times in the upcoming year. Once the basic story was in place, they knew they could make the minor changes needed to make it fit the various venues.

The biggest drawback of outsourcing is that you don't develop the skills and learn in the same way as if you tackle the challenge internally. To help overcome this, look for opportunities to learn from the consultant during the process. Consider whether the output can also provide a starting point for other work, or if it can be evolved over time as you develop internal capability.

A combined approach

The teams and organizations I've seen become the most successful in this space leverage a combined approach. They recognize the importance of storytelling with data and invest in training and practice to give everyone the foundational knowledge for effective data visualization. They also identify and support an internal expert, to whom the rest of the team can turn for help overcoming specific challenges. They bring in external experts to learn from as makes sense. They recognize the value of being able to tell stories with data effectively and invest in their people to build this competency.

Through this book, I've given *you* the foundational knowledge and language to use to help your team and your organization excel when it comes to communicating with data. Think about how you can frame feedback in terms of the lessons we've covered to help others improve their ability and effectiveness as well.

Let's wrap up with a recap of the path we've taken to effective storytelling with data.

Recap: a quick look at all we've learned

We have learned a great deal over the course of this book, from context to cutting clutter and drawing attention to telling a robust story. We've worn our designer hats and looked at things through our audience's eyes. Here is a review of the main lessons we've covered:

1. **Understand the context.** Build a clear understanding of who you are communicating to, what you need them to know or do, how you will communicate to them, and what data you have to back up your case. Employ concepts like the 3-minute story, the Big Idea, and storyboarding to articulate your story and plan the desired content and flow.

2. **Choose an appropriate visual display.** When highlighting a number or two, simple text is best. Line charts are usually best for continuous data. Bar charts work great for categorical data and must have a zero baseline. Let the relationship you want to show guide the type of chart you choose. Avoid pies, donuts, 3D, and secondary *y*-axes due to difficulty of visual interpretation.

3. **Eliminate clutter.** Identify elements that don't add informative value and remove them from your visuals. Leverage the Gestalt principles to understand how people see and identify candidates for elimination. Use contrast strategically. Employ alignment of elements and maintain white space to help make the interpretation of your visuals a comfortable experience for your audience.

4. **Focus attention where you want it.** Employ the power of pre-attentive attributes like color, size, and position to signal what's important. Use these strategic attributes to draw attention to where you want your audience to look and guide your audience through your visual. Evaluate the effectiveness of preattentive attributes in your visual by applying the "where are your eyes drawn?" test.

5. **Think like a designer.** Offer your audience visual affordances as cues for how to interact with your communication: highlight the important stuff, eliminate distractions, and create a visual hierarchy of information. Make your designs accessible by not over-complicating and leveraging text to label and explain. Increase your audience's tolerance of design issues by making your visuals aesthetically pleasing. Work to gain audience acceptance of your visual designs.

6. **Tell a story.** Craft a story with clear beginning (plot), middle (twists), and end (call to action). Leverage conflict and tension to grab and maintain your audience's attention. Consider the order and manner of your narrative. Utilize the power of repetition to help your stories stick. Employ tactics like vertical and horizontal logic, reverse storyboarding, and seeking a fresh perspective to ensure that your story comes across clearly in your communication.

Together, these lessons set you up for success when communicating with data.

In closing

When you opened this book, if you felt any sense of discomfort or lack of expertise when it comes to communicating with data, my hope is that those feelings have been mitigated. You now have a solid foundation, examples to emulate, and concrete steps to take to overcome the data visualization challenges you face. You have a new perspective. You will never look at data visualization the same. You are ready to assist me with my goal of ridding the world of ineffective graphs.

There is a story in your data. If you weren't convinced of that before our journey together, I hope you are now. Use the lessons we've covered to make that story clear to your audience. Help drive better decision making and motivate your audience to act. Never again will you simply show data. Rather, you will create visualizations that are thoughtfully designed to impart information and incite action.

Go forth and tell your stories with data!

bibliography

Arheim, Rudolf. *Visual Thinking*. Berkeley, CA: University of California Press, 2004.

Atkinson, Cliff. *Beyond Bullet Points: Using Microsoft PowerPoint to Create Presentations that Inform, Motivate, and Inspire*. Redmond, WA: Microsoft Press, 2011.

Bryant, Adam. "Google's Quest to Build a Better Boss." *New York Times*, March 13, 2011.

Cairo, Alberto. *The Functional Art: An Introduction to Information Graphics and Visualization*. Berkeley, CA: New Riders, 2013.

Cohn, D'Vera, Gretchen Livingston, and Wendy Wang. "After Decades of Decline, a Rise in Stay-at-Home Mothers." *Pew Research Center*, April 8, 2014.

Cowan, Nelson. "The Magical Number Four in Short-Term Memory: A Reconsideration of Mental Storage Capacity." *Behavioral and Brain Sciences* 24 (2001): 87–114.

Duarte, Nancy. *Resonate: Present Visual Stories that Transform Audiences*. Hoboken, NJ: John Wiley & Sons, 2010.

Duarte, Nancy. *Slide:ology: The Art and Science of Creating Great Presentations*. Sebastopol, CA: O'Reilly, 2008.

Few, Stephen. *Show Me the Numbers: Designing Tables and Graphs to Enlighten*. Oakland, CA: Analytics Press, 2004.

Few, Stephen. *Now You See It: Simple Visualization Techniques for Quantitative Analysis*. Oakland, CA: Analytics Press, 2009.

Fryer, Bronwyn. "Storytelling that Moves People." *Harvard Business Review*, June 2003.

Garvin, David A., Alison Berkley Wagonfeld, and Liz Kind. "Google's Project Oxygen: Do Managers Matter?" Case Study 9–313–110, *Harvard Business Review*, April 3, 2013.

Goodman, Andy. *Storytelling as Best Practice*, 6th edition. Los Angeles, CA: The Goodman Center, 2013.

Grimm, Jacob, and Wilhelm Grimm. *Grimms' Fairy Tales*. New York, NY: Grosset & Dunlap, 1986.

Iliinsky, Noah, and Julie Steele. *Designing Data Visualizations*. Sebastopol, CA: O'Reilly, 2011.

Klanten, Robert, Sven Ehmann, and Floyd Schulze. *Visual Storytelling: Inspiring a New Visual Language*. Berlin, Germany: Gestalten, 2011.

Lidwell, William, Kritina Holden, and Jill Butler. *Universal Principles of Design*. Beverly, MA: Rockport Publishers, 2010.

McCandless, David. *The Visual Miscellaneum: A Colorful Guide to the World's Most Consequential Trivia*. New York, NY: Harper Design, 2012.

Meirelles, Isabel. *Design for Information*. Beverly, MA: Rockport Publishers, 2013.

Miller, G. A. "The Magical Number Seven, Plus or Minus Two: Some Limits on Our Capacity for Processing Information." *The Psychological Review* 63 (1956): 81–97.

Norman, Donald A. *The Design of Everyday Things*. New York, NY: Basic Books, 1988.

Reynolds, Garr. *Presentation Zen: Simple Ideas on Presentation Design and Delivery*. Berkeley, CA: New Riders, 2008.

Robbins, Naomi. *Creating More Effective Graphs*. Wayne, NJ: Chart House, 2013.

Saint-Exupery, Antoine de. *The Airman's Odyssey*. New York, NY: Harcourt, 1943.

Simmons, Annette. *The Story Factor: Inspiration, Influence, and Persuasion through the Art of Storytelling*. Cambridge, MA: Basic Books, 2006.

Song, Hyunjin, and Norbert Schwarz. "If It's Hard to Read, It's Hard to Do: Processing Fluency Affects Effort Prediction and Motivation." *Psychological Science* 19 (10) (2008): 986–998.

Steele, Julie, and Noah Iliinsky. *Beautiful Visualization: Looking at Data Through the Eyes of Experts*. Sebastopol, CA: O'Reilly, 2010.

Tufte, Edward. *Beautiful Evidence*. Cheshire, CT: Graphics Press, 2006.

Tufte, Edward. *Envisioning Information*. Cheshire, CT: Graphics Press, 1990.

Tufte, Edward. *The Visual Display of Quantitative Information*. Cheshire, CT: Graphics Press, 2001.

Tufte, Edward. *Visual Explanations: Images and Quantities, Evidence and Narrative*. Cheshire, CT: Graphics Press, 1997.

Vonnegut, Kurt. "How to Write with Style." *IEEE Transactions on Professional Communication* PC-24 (2) (June 1985): 66–67.

Ware, Colin. *Information Visualization: Perception for Design*. San Francisco, CA: Morgan Kaufmann, 2004.

Ware, Colin. *Visual Thinking for Design*. Burlington, MA: Morgan Kaufmann, 2008.

Weinschenk, Susan. *100 Things Every Designer Needs to Know about People*. Berkeley, CA: New Riders, 2011.

Wigdor, Daniel, and Ravin Balakrishnan. "Empirical Investigation into the Effect of Orientation on Text Readability in Tabletop Displays." Department of Computer Science, University of Toronto, 2005.

Wong, Dona. *The Wall Street Journal Guide to Information Graphics*. New York, NY: W. W. Norton & Company, 2010.

Yau, Nathan. *Data Points: Visualization that Means Something*. Indianapolis, IN: John Wiley & Sons, 2013.

Yau, Nathan. *Visualize This: The FlowingData Guide to Design, Visualization, and Statistics*. Indianapolis, IN: John Wiley & Sons, 2011.

Index

A

Accessibility, 138–145, 198
  overcomplicating, 139–141
  poor design, 139
  text, thoughtful use of,
    141–145
    action titles on slides, 141
Action words, 23
Adobe Illustrator, 244
Aesthetics, 145–148, 198
Affordances, 128–138
  creating a clear visual
    hierarchy of information,
    135–138
  eliminating distractions,
    132–135
  highlighting effects, 129–132
Alignment, 82–84
  diagonal components, 83–84
  presentation software tips for,
    83
Animation, leveraging in
    visuals, 210–218
Annotated line graph with
    forecast, 154
Area graphs, 59–60
Atkinson, Cliff, 172

Audience attention, focusing,
    15, 99–126, 195–197
  color, 117–124
    brand colors, 123–124
    considering tone
      conveyed, 122–123
    designing with colorblind
      in mind, 121–122
    position on page, 124–126
    using consistently, 120–121
    using sparingly, 118–120
  memory, 100–102
    iconic, 101
    long-term, 102
    short-term, 101–102
  preattentive attributes,
    102–116
    in graphs, 109–116
    in text, 106–109
  sight, 100
  size, 116–117

B

Bar charts, 50–59, 156–158,
    161–162, 236–237
  axis vs. data labels, 52
  bar width, 53

Bar charts (*continued*)
  categories, logical ordering
    of, 58
  ethical concerns, 53
  horizontal, 57
  simple, 236–237
  stacked
    horizontal, 58–59, 161–162,
      237–238
    leveraging positive and
      negative, 158
    100%, 156–158
    vertical, 54–55
  vertical, 54
  waterfall chart, 55–57
Beck, Harry, 139
Beyond Bullet Points (Atkinson),
  172
Big Idea, 30–31, 189
Bing, Bang, Bongo, 180–181

C
Cairo, Alberto, 248
Case studies, 17, 207–240
  alternatives to pie charts,
    234–240
    100% stacked horizontal
      bar graph, 237–238
    showing numbers directly,
      236
    simple bar graph, 236–237
    slopegraph, 238–240
  color considerations with a
    dark background,
    208–210
  animation, leveraging in
    visuals, 210–218
  logic in order, 219–227

  spaghetti graphs, avoiding,
    227–234
    combined approach,
      232–234
    emphasizing one line at a
      time, 229–230
    separating spatially,
      230–232
Closure principle, 78, 92
Clutter, avoiding, 15, 71–98
  cognitive load, 71–73
    data-ink/signal-to-noise
      ratio, 72
  contrast, nonstrategic use of,
    86–90
    redundant details, use of,
      90
  decluttering, 90–97
    cleaning up axis labels, 95
    labeling data directly, 96
    leveraging consistent color,
      97
    removing chart border, 92
    removing data markers,
      94
    removing gridlines, 93
  Gestalt Principles of Visual
    Perception, 74–81
    closure, 78, 92
    connection, 80
    continuity, 79
    enclosure, 77
    proximity, 75, 96
    similarity, 76, 97
  presence of, 73
  visual order, lack of, 81–86
    alignment, 82–84
    white space, 84–86

Cognitive load, 71–73
    data-ink/signal-to-noise ratio,
        72
Color considerations with a
        dark background,
        208–210
Color saturation, 42
Communication mechanism
        continuum, 24
    live presentation, 24–25
    slideument, 26
    written document or email,
        25–26
Connection principle, 80
Context, importance of, 14,
        19–33, 188–189
    Big Idea, 30–31
    consulting for, 28–29
    exploratory vs. explanatory
        analysis, 19–20
    how, 26
    illustrated by example,
        27–28
    supporting data, 27
    storyboarding, 31–33
    3-minute story, 30
    understanding, 188–189
    what, 22–26
        action, 22–23
        mechanism, 23–26
        tone, 26
    who, 21–22
        audience, 21
        you, 21–22
Continuity principle, 79
Contrast, nonstrategic use of,
        86–90
    redundant details, use of, 90

D
Data-ink ratio, 72
Data Points (Yau), 20
Distractions, eliminating,
        132–135
Donut charts, 65
Duarte, Nancy, 22, 30, 72, 173
        179

E
Eager Eyes (blog), 247
Effective visuals, choosing, 14,
        35–69
    graphs, 43–49
        area graphs, 59–60
        bar charts, 50–59
        lines, 45–49
        points, 44–45
        slopegraph, 47–49
    infographics, 60–61
    simple text, 38–40
    tables, 40–43
        borders, 41
        heatmap, 42–43
    visuals to avoid, 61–68
        3D charts, 65
        donut charts, 65
        pie charts, 61–65
        secondary y-axis,
            66–67
Enclosure principle, 77
Excel, 13, 42, 244
    changing components of a
        graph in, 196
    slopegraph template, 48
Exploratory vs. explanatory
        analysis, 19–20,
        112

F

Few, Stephen, 41, 105, 248
FiveThirtyEight's Data Lab, 247
Flowing Data (blog), 247
The Functional Art (blog), 248
Fung, Kaiser, 248

G

Gestalt Principles of Visual
        Perception, 74–81
    closure, 78, 92
    connection, 80
    continuity, 79
    enclosure, 77
    proximity, 75, 96
    similarity, 76, 97
Google
    People Analytics, 9–10
    Project Oxygen, 10
    spreadsheets, 243
Graphs, 43–49
    area graphs, 59–60
    bar charts, 50–59
        axis vs. data labels, 52
        bar width, 53
        categories, logical ordering
            of, 58
        ethical concerns, 53
        horizontal, 57
        stacked horizontal, 58–59
        stacked vertical, 54–55
        vertical, 54
        waterfall chart, 55–57
    lines, 45–49
        line graph, 46–47
    points, 44–45
        scatterplots, 44–45
    slopegraphs, 47–49
        modified, 49
        template, 48
The Guardian Data Blog, 248

H

Headlines, creating, 174
Heatmap, 42–43
HelpMeViz (blog), 248
Hierarchy of information,
        135–138
    super-categories, 136
Highlighting effects, 129–132
Horizontal logic, 181–182
"How to Write with Style"
        (Vonnegut), 170

I

Iconic memory, 101
Ineffective graphs, examples
        of, 1
Infographics, 60–61
Information Visualization:
        Perception for Design
        (Ware), 86

K

Kirk, Andy, 248
Kriebel, Andy, 248

L

Line graph, 46–47, 152–154
    annotated with forecast, 154
Live presentation, 24–25
    tables in, 40
Logic in order, 219–227
Long-term memory, 102, 179

M

Make a Powerful Point (blog), 248
McCandless, David, 123, 142
McKee, Robert, 168
McMahon, Gavin, 248
Model visuals, dissecting, 16, 151–163
   line graph, 152–154
     annotated with forecast, 154
     stacked bars
       horizontal, 161–162
       leveraging positive and negative, 158
       100%, 156–158
Moonville example, 211–218

P

Perceptual Edge (blog), 248
Pie charts, 61–65, 235
Points, 44–45
   scatterplots, 44–45
PowerPoint, 244
Preattentive attributes, 102–116
   in graphs, 109–116
   in text, 106–109
Proximity principle, 75, 96

R

Resonate (Duarte), 22, 30, 72
Reverse storyboarding, 183

S

Scatterplots, 44–45
   modified, 45
Schwabish, Jon, 248

Secondary y-axis, 66–67
Short-term memory, 101–102
Show Me the Numbers (Few), 41
Signal-to-noise ratio, 72
Similarity principle, 76, 97
Simple text, 36, 38–40
Slideument, 26, 211
Slopegraphs, 47–49, 238–240
   modified, 49
   template, 48
Spaghetti graphs, avoiding, 227–234
   combined approach, 232–234
   emphasizing one line at a time, 229–230
   separating spatially, 230–232
Spears, Libby, 168
Stacked bars
   horizontal, 161–162
   leveraging positive and negative, 158
   100%, 156–158
Storyboarding, 31–33
Storytelling, 16, 165–185
   constructing the story, 171–174
     beginning, 171–173
     end, 174
     middle, 173–174
   lessons in, 16
   magic of story, 166–171
     in cinema, 168–170
     in plays, 167–168
     in written word, 170–171
   narrative structure, 175–179
     narrative flow, 175

Storytelling (*continued*)
    spoken and written,
        177–179
    repetition, 179–181
        Bing, Bang, Bongo,
            180–181
    tactics to ensure the story is
        clear, 181–184
        horizontal logic, 181–182
        reverse storyboarding, 183
        vertical logic, 182–183
storytelling with data (blog),
    248
Storytelling with data process,
    187–205, 242–255
    appropriate display,
        choosing, 189–193, 253
    audience attention, focusing,
        195–197, 254
    building competency in team
        or organization, 250–253
        combined approach,
            252–253
        investing in internal
            experts, 252
        outsourcing, 252
        upskilling everyone, 251
    clutter, eliminating, 193–194,
        254
    context, understanding,
        188–189, 253
    telling a story, 199–204, 254
    thinking like a designer,
        197–198, 254
    tips for success with, 242–255
        devoting time to, 246–247
        having fun and finding your
            style, 249–250

        iterating and seeking
            feedback, 245–246
        seeking inspiration through
            good examples, 247–249
        tools, learning to use,
            243–245
Super-categories, 136, 137
Survey feedback, 59, 81, 209, 219

T
Tableau, 243–244
Tables, 40–43
    borders, 41
    heatmap, 42–43
Thinking like a designer, 15–16,
    127–150
    acceptance, 149–150
    accessibility, 138–145
        overcomplicating, 139–141
        poor design, 139
        text, thoughtful use of,
            141–145
    aesthetics, 145–148
    affordances, 128–138
        creating a clear visual
            hierarchy of information,
            135–138
        eliminating distractions,
            132–135
        highlighting effects, 129–132
3-minute story, 30
3D charts, 65
Tufte, Edward, ix, 72, 231

U
Universal Principles of Design
    (Lidwell, Holden, and
    Butler), 129, 149

V

Vertical logic, 182–183
The Visual Display of
        Quantitative Information
        (Tufte), 72
The Visual Miscellaneum;
        A Colorful Guide
        to the World's Most
        Consequential Trivia
        (McCandless), 123
Visual order, lack of, 81–86
    alignment, 82–84
        diagonal components,
        83–84
        presentation software tips
        for, 83
    white space, 84–86
Visualising Data (blog), 248
Visuals to avoid, 61–68

3D charts, 65
donut charts, 65
pie charts, 61–65
secondary y-axis, 66–67
VizWiz (blog), 248
Vonnegut, Kurt, 170

W

Ware, Colin, 86, 118
Waterfall chart, 55–57
    brute-force, 56–57
White space, 84–86
Written document or email,
        25–26
WTF Visualizations (wtfviz.net),
        249

Y

Yau, Nathan, 20, 247